THE
SUPERNATURAL
SCIENCE
THEORY AND MAGIC

DAVID BARRETO

Copyright© 2019 by David Barreto
All rights reserved.

THE
SUPERNATURAL
SCIENCE
THEORY AND MAGIC

First edition: 2019, London, United Kingdom

ISBN 978-1-9162111-4-8

Cover and design by David Barreto

www.davidbarreto.net

No part of this book may be reproduced or used in any form or by any means, including electronic or mechanical methods such as photocopying, recording, or as a downloadable file, without the prior written permission of the author.

AUTHOR NAME

THE SUPERNATURAL SCIENCE
THEORY AND MAGIC

Contents

Preface i

Introduction 1

PART 1 God and The Spiritual Hierarchies 6

    God 7

    Solar Logos and Planetary Christ 13

    Angels 16

    Spirit Guides 17

    Devas and Elementals 19

    Karma Agents 22

    Spiritual Vampires 27

    Ghosts and Poltergeists 30

PART 2 Mediumship and Psychic Perception 37

    The Chakras and the Three Auras 39

    Mediumship 46

    Psychic Perception 53

    Dreams and Astral Travel 62

    Meditation and the Unlocking of Potentials 70

PART 3 Low Magic and Ritual Practices 75

    Witchcraft Spells, Voodoo, and Incantation 76

    Pacts and Spiritual Contracts 85

    Offerings 87

    Prayer and Idols 92

PART 4 Divination, Oracles and Talismans..........................97

    Magic Symbols..98

    Oracular Tools..103

    Palmistry..107

    Supernatural Games..110

Bibliography..113

Recommended Readings..119

PREFACE

This book contains terms that may be unfamiliar to some readers, such as "astral," "etheric," "semi-physical," and "electromagnetic." The term "astral" refers to something related to the spiritual planes that is not physical or bound to the material world. "Etheric" describes something subtle, such as an element, fluid, or material, that is too refined to be considered physical yet not entirely spiritual. "Semi-physical" refers to something that has both physical and spiritual attributes. "Electromagnetic" and "electromagnetism" describe the movement of charged particles or waves through fields, and while they are primarily used to describe physical objects or bodies, they can also serve as an analogy for extraphysical phenomena.

Throughout the book, the reader may encounter discussions on the different bodies of an individual, such as the astral body, the lower mental body, the buddhic body, and so on. These bodies should be viewed as vessels for a more subtle body to inhabit, allowing for the experience of reality at a lower frequency or denser dimension. The physical body serves as the vessel for the astral body, which in turn serves as the vessel for the lower and upper mental bodies. The latter two bodies, in turn, serve as the vessel for the buddhic body and the atman. The physical body translates spiritual movements into the physical universe, while the astral body decodes one's emotions. The lower mental body decodes the five senses and the intellect, whereas the upper mental body is responsible for one's will. The buddhic body encompasses one's divine consciousness, and the atman is a fragment detached from the source, on a journey to acquire its own merits and expand creation.

Most parts of this work will discuss the aura and its various layers, including the physical, semi-physical, and spiritual auras. The aura is a set of energetic layers of radiation or emanation that surround a body, and each layer is composed of the same type of energy as the main center it revolves around, depending on the level of condensation and frequency it resides in.

The book may mention life outside Earth, but it is important to note that these passages are referring to life on another plane or dimension, not on the physical planet of the material universe. Therefore, if the book mentions something along the lines of "reincarnating in Jupiter," it is essentially referring to the astral planes of Jupiter.

The reader will often encounter words such as "energy," "vibration," and "frequency," which are commonly used in spiritual literature and may sound ambiguous to some. In this context, "energy" refers to the movement of light, "vibration" is the movement of something, and "frequency" is how this movement occurs. For example, in the phrase "the energy of the crown chakra vibrates at a high frequency," it can be understood as "the moving light of the crown chakra moves rapidly."

It is also possible that some readers may not be familiar with certain terms, such as "discarnate" and "incarnate." In the context of this book, "incarnate" refers to an entity that possesses a physical body, while "discarnate" refers to an entity without a physical body, often in reference to those who have passed away. "Incarnation" represents the act of entering into a physical form for a particular lifetime, while "reincarnation" involves the process of re-entering a physical form after having done so in previous lives.

Additionally, this book employs terminology commonly used in the fields of physics, chemistry, and quantum mechanics for instructional purposes. While there may be similarities between these sciences and theories, it is important to note that the use of these terms is not intended to imply a direct relationship or equivalence.

Based on the current understanding of these terms, readers can expect to encounter the following in the book: a foundational understanding of the concept of God as the first topic, striving to expound on what the human mind can comprehend. Unfortunately, the term "god" is often misunderstood, leading many students of esotericism to disregard it altogether due to disappointments and unconscious religious traumas. However, it is crucial to note that the term "god" does not necessarily need replacement but rather a shift in perception. The depiction of God in Abrahamic religions, for example, may be limited to a particular personality and appearance, but this does not justify rejecting the existence of God entirely, as though God could only be that God or none at all. My inquiry extends beyond ancient religions or scriptures, as I do not address a god from a specific religion but instead focus on an all-encompassing God who is not a singular entity or spirit. After discussing God, the book proceeds to explore the nature and roles of entities such as angels, spirit guides, and nature spirits in a universal and lucid manner as part of the divine and spiritual hierarchies. Essentially, this chapter aims to clarify common misunderstandings about these groups and offer insight into who those spirits are and why they exist. This part also focuses on the karmic influence of spirits, besides thoroughly dissecting the idea of ghosts and poltergeist, providing readers with a comprehensive understanding of all entities, from the highest to the lowest.

The second part of the book is devoted to a comprehensive and impartial analysis of various types of mediumship and paranormal phenomena. As a significant portion of the book is based on mediumship and psychic experiences, I aim to identify the hidden explanations for these events, elucidate the physical and spiritual mechanisms that unfold during these events, and clarify why only some people possess these abilities and why they are not universally experienced. Moreover, I will offer a detailed description of the involvement of spirits in these occasions. I will also provide a sober explanation on astral traveling, and how meditation can enhance one's psychic and mediumship experiences.

In the third part of the book, I delve into the practices of witchcraft, lower magic, and enchantment. Through my analysis, I aim to provide a profound interpretation of the mechanisms behind spells and incantations, uncovering what lies hidden beneath these practices. I will also explore how spells progress in the invisible realms, with a particular focus on the potential involvement of spirits. In this chapter, I aim to elucidate how and why spells are created in specific ways, detailing descriptions of the energies used, including their composition, creation process, and methods of dismantlement. This chapter offers readers the opportunity to understand the occult nature of offerings given to spirits and entities, including insight into how they use and interact with these offerings. Furthermore, this chapter provides insights into the nature of prayer: how deities may receive it, whether they respond to it, and how they might proceed if they do.

The final section of the book delves into the world of oracles and talismans, offering insights into how divinatory tools function, the occult role of protection charms, and the science behind tarot readings. It also provides detailed explanations of how palmistry works, aura and energy reading, and the spiritual dangers and misconceptions of ouija boards.

In writing this book, my aim is to reveal mysteries and dispel myths about the supernatural and the spiritual universe. I have no intention of converting skeptics into believers or believers into skeptics, but rather to provide readers who are interested in these topics with a deeper, more universal understanding of them, presented in a way that is both accessible and grounded in scientific thinking.

INTRODUCTION

The realm of the unknown and the occult, which is simply a universe that has not been thoroughly investigated, often includes esoteric or spiritual practices and phenomena that are not necessarily unknown but rather misunderstood. In order to shed light on these topics, I have explored various subjects through a scientific lens and found that they can be elucidated with greater clarity. In my opinion, those who seek a hyper-mystical explanation of these themes may actually encounter superficiality because they tend to judge a book by its cover or a speaker by their appearance and age. An older man wearing a turban may be perceived as having more authority and appealing to curious learners than a younger man in modern attire and a trendy hairstyle. In fact, a spiritual teacher I admire once conducted an experiment in which he asked his students to describe a picture of a bearded man wearing a turban projected on a screen. Most of them assumed that the man was an Indian guru, a yogi master, or a sage, when in reality he was a homeless person taking a bath in a fountain in Mumbai. It is important to recognize that while a person's physical age may contribute to their wisdom, the spirit that governs their body is ageless, and therefore a person's physical age does not necessarily reflect the maturity of their spirit. It is unfortunate that stereotypes about the appearance of those who are believed to hold transcendental truths still persist.

Likewise, persistent stereotypes surrounding practices like witchcraft and beliefs in ghosts can create a dismissive attitude toward these topics, hindering their recognition as serious subjects of study. Another common misconception among people is that mediums and psychics possess inherent supernatural abilities and are more enlightened than most individuals. However, like everyone else, these individuals are lifelong learners and are, like everyone else, subject to the laws of cause and effect. If they possess these abilities, it is not because they were born to be more capable than others, but because they have a responsibility to use their abilities to serve others without seeking personal gain. In fact, their condition can be more of a burden than a joyful gift. It is also important to recognize that mediums, psychics, and others who work with spiritual practices are ordinary people who face their own share of difficulties, losses, and hardships. They are not exempt from illness, accidents, or emotional pain, and it is unrealistic to expect them to always appear serene, happy, healthy, and beautiful. It is not their role to predict lottery numbers or converse with high-ranking spirits at will; nor are they meant to live a perfect life or live beyond the age of 120 years.

One more frequent misunderstanding is that simply reading several books on spirituality leads to enlightenment, but true spiritual growth comes from embodying qualities such as altruism, kindness, and benevolence in one's daily life. While knowledge is certainly important for spiritual development, some individuals mistakenly believe that they are superior to others due to their supposed knowledge or degree collection. They may even believe that they will be granted a special status after death, but in reality, we are all here to experience difficult lessons and together grow our sense of fraternity. Being a medium, lecturer, or author on spirituality does not automatically qualify one for angelic status. My own journey in spirituality has taught me that true growth comes not from theoretical knowledge but from the consistent practice of virtuous qualities. Despite my deep interest in astrology, tarot, and books from crystal shops since my preteen years, I now realize that these interests are tools to aid in my spiritual development, but not the ladder.

As a universalist, I have studied from every source, religion, and niche and have visited spiritualist centers to witness mediums channeling spirits and conveying information they could not possibly have known. I have also visited various places of worship, including temples of folk religion where both mediumship and psychic ceremonies were held. It was in these places that I met teachers who imparted knowledge that I could not easily glean from books, helping me develop my own way of connecting to the spiritual realm. As an individual with a deep interest in the mysteries of the universe, I have also pursued the study of both astrophysics and religious scriptures in order to gain a better understanding of the cosmos and the insights that ancient cultures have to offer on my favorite subjects.

Along the way of my scientific and esoteric studies, I came across intriguing questions about the extraphysical realm: do oracles really work, and if so, what force is behind them? If ghosts exist, what are they made of? How do telepaths read our minds, and how do mediums hear spirits when others cannot? As I delved deeper into these subjects, I began to seek answers through deep meditation and further study. Through this process, I developed a strong conviction that the information contained within this book was revealed to me through a mysterious and serendipitous process. As my understanding deepened, I found that the knowledge I had somehow acquired was aligned with what I found through rigorous research and investigation. But as I searched for more knowledge, I felt the need for a solid foundation to support my newfound understanding. As a skeptic, I began exploring the intersection between esoteric learning and worldly science.

Science has undoubtedly played a significant role in the progress of the world, but unfortunately, it often serves the interests of financial tyranny. While technology has given us access to high-tech discoveries that have made our lives easier, it seems to be failing at elevating our morals. Despite boasting of scientific advancements, the world still faces significant challenges such as poverty and exploitation, which could be eradicated with the resources dedicated to space tourism and other superficial technological pursuits that aim to benefit dozens as opposed to billions.

As I delved deeper into Cartesian science, I initially struggled to reconcile its discoveries with spirituality, as research on the "occult" is rarely funded. I was disappointed to find that phenomena such as the power of the mind or spiritual healing are often dismissed as unreal, a religious fallacy, or pseudoscientific. But I also believe that as technology advances, materialist science gains a deeper understanding of the natural world, and what was once considered supernatural can rapidly be redefined as a physical phenomenon. I would even dare to say that if technological sensors capture semi-physical auras, they may lead scientists to describe them as "ultra-electrical cellular currents of human metabolism." Overall, I realized that science heavily relies on technology to augment our senses, such as microscopes, antennas, and sensors, allowing us to see, hear, and feel what may be otherwise beyond our natural capabilities. So I concluded that the storm of dismissals is, sadly, due to our primitive technology. Science is a discipline that belongs to no individual or group. Rather, it is open for all who wish to engage in the inquiry of the universe through the application of scientific methods or critical thinking.

Despite this, we should bear in mind the wise words of Carl Sagan, who famously said that "the absence of evidence is not evidence of absence."

PART 1

GOD AND THE SPIRITUAL HIERARCHIES

God

To begin this work, it is essential to provide an explanation of the concept of 'God.' This necessity arises from the fact that the subject matter discussed in this book pertains to the creations of God. However, it is noteworthy that the human capacity to comprehend God is limited to its physical aspects. The spiritual and divine totality of God is beyond observational evidence and exceeds the capacity of the human brain to fully comprehend. Even on the spiritual planes, humans may not be able to fully comprehend God's immensity. Therefore, the explanation presented in this book will focus on the observational creation of God, and the explanation of God may be intertwined with an explanation of its creations.

The physical brain serves as a reducing valve for consciousness, allowing the spirit to immerse itself in an incarnated life and experience materialistic constraints. A discarnate spirit may experience broader perceptions and relative freedom, depending on its level of enlightenment. However, upon incarnating, it finds that its physical brain is merely a rudimentary reality decoder, enabling sensory cells to detect or translate physical phenomena. According to Von Barthheld (2016), the human brain consists of approximately 90 billion neurons, most of which are only capable of processing sensory and physiological information. When a sound is heard, an image is seen, or a smell is recognized, the brain processes and decodes that information solely based on its material nature. Hence, the brain can be perceived as a reader of physical phenomena, translating vibrations and molecules into detectable reality. Anything that cannot be touched, smelled, heard, tasted, or seen by physical sensory organs is readily dismissed or assumed to be imagination.

Most extra-physical phenomena captured by the physical brain are conducted via sensory association. This means that the information is cataloged as a physical phenomenon that matches what the physical brain has already experienced. As a result, any explanation of God depends on the physical brain, which attempts to understand it in an observational manner as though it were a material entity. This fact leads to the assumption that to fully understand the concept of God, one must evolve spiritually to a level where physical experiences are no longer the only parameter used to discern something beyond physical explanation. Fully comprehending God is an unattainable feat for any human being. However, gaining an understanding of its physical essence may serve as a preliminary step. Skeptics and Cartesian thinkers may assert that the existence of a deity cannot be proven. Nevertheless, attempts to demonstrate God's existence through a microscopic lens are futile, for God, though its creator, is not a material entity. Employing materialistic theories to establish the presence or absence of a non-physical entity is unproductive and devoid of meaning. Ultimately, the confirmation or negation of God's existence does not necessitate academic degrees.

God is not a person, a spirit, or a being, but rather a flawless structure that encompasses all that exists. This includes strings, quarks, and leptons; atoms; elements; planets; galaxies; fundamental forces; quantum fields; intelligence; and time, all of which are embodiments of God. In traditionally Christian nations, God is commonly regarded as the Abrahamic almighty who created the heavens and the earth in seven days; formed Adam and Eve from dust; and is the direct father of Jesus. However, this description stems from a specific religion and does not imply that God can only be defined in this manner. The Abrahamic God may be characterized by gender, age, appearance, and a location from where he governs, but these are qualities attributed by the initial prophets and disseminators of knowledge to describe a system that was omniscient, omnipotent, and omnipresent. They most likely employed the image of an old man on a throne in the heavens as a means of conveying that the system was intelligent, powerful, and universally present. Similarly, the gods of many other cultures were given gender, physical attributes, and human personality traits as a way of conveying a message that ancient peoples could readily comprehend. The concept of God, or any god, refers to a supreme entity, or system, that acts as a creator, sustainer, or ruler of something. For example, a specific god may be viewed as the creator of the universe, the ruler of the seas, or the patron of love. Along these lines, the notion of a singular god encompasses the roles of creator, sustainer, and ruler of all existence, including both physical and non-physical reality.

The big bang theory is commonly associated with the origin of "everything," but it specifically pertains to the beginning of the physical universe, physical space, and physical time. This theory was first introduced in 1927 by Georges Lemaître, a Belgian astronomer and priest. Lemaître proposed that the universe emerged from a single point, which he referred to as the "primeval atom," and then it underwent a rapid expansion that produced substantial amounts of matter and energy. Although the big bang was the first scientific theory to be promulgated, ancient religious literature, such as the hymns of the Rig Vedas, also mentions the theory (Moorthy, 2011). If the big bang theory is true, which it is likely to be, it means that the physical universe and, by extension, life began in a more condensed planes. In this sense, God, as the cause of everything, expanded its own existence. But nevertheless, using concepts of quantum mechanics and modern physics may only reveal the "physical properties" of God, that is, its creations, and still that would not be appropriately sensible, as God is present in the divine planes as divinity, in the spiritual planes as spiritual energy, and in the physical planes as material creation.

Discussing the creations of God without considering dimensions would be an error, as they exist in various planes of reality. In physics, dimension refers to the space defined by coordinates. A dimension, also referred to as a "plane" in esoteric and spiritual literature, denotes a location that may overlap with one or more other locations. Rather than being similar to the layers of an onion, they are more akin to different radio waves penetrating the same space without interfering with each other, as they reside at different frequencies. In this context, frequency may be the most accurate and closest explanation of the existence of other planes. If God is divine, spiritual, and material, there are at least three planes that can be explored. In each of these planes, multiple layers or frequencies exist, implying that creation permeates and occupies all spaces. Therefore, when it is said that an individual departs their physical body after death and enters another plane or dimension, it implies that the individual is now in a particular plane that is most compatible with their most dense body; in other words, as they have left their physical body, it is natural for them to no longer experience the physical dimension.

Overall, God is movement, and as stated by Nikola Tesla, "If you want to find the secrets of the universe, think in terms of energy, frequency, and vibration", one may grasp that all the aforementioned qualities of movement can be attributed to being natural to God. Energy, frequency, and vibration certainly connote movement. The big bang theory suggests that physical reality emerged from some sort of movement, and modern physics heavily describes the movement and oscillations of the numerous fields in which physical creation is still taking place. Some of these fields comprehend, among others, the Unified Field Theory of Gravitation and Electricity, theorized and assiduously defended by Einstein (1925). As asserted by Feynman et al. (1965), that field, as well as any other field, is depicted by oscillations. In such a way, it can be presumed that a change in vibration in any part of the space-time locale gives rise to particles, whatever the field they pertain to.

Under spiritualist views, vibrations that give rise to objects can be associated with an emotion or a circumstance that humans may describe as: faith, knowledge, law, love, justice, generation, evolution, and many subdivisions thereof. The divine vibration that assembles an apple, for instance, is analogous in frequency to love. Based on the frequency of love observed in humans, a similar aura is observed in the physical aura of an apple. The generation of an apple refers to the movement of particles and, subsequently, of elements that come to form it. It is not that an apple means love; it is that apples are expressions of love in the realm of flora, and more specifically, in the group of fruits. Along these lines, the perception of one or more divine figures, not necessarily representing any specific religion, may be legitimately correct in both monotheistic and polytheistic cultures. A single god can be subdivided into several others, depending on how "it" is empirically understood. If the most approximate description of God's physical expression is vibration and frequency, then one vibration can conveniently be dissected into various frequencies. An only god may be compared to sunlight: after it is reflected, refracted, and dispersed, the electromagnetic radiation is perceived by the human eye as an array of colors, just like a rainbow. Thus, light is not colorless but white, which is the combination of all the colors in the spectrum. In polytheistic cultures, each frequency of God was shown by a different god, just like the different colors in a rainbow show the different frequencies of sunlight. As previously stated, God exists in the divine, spiritual, and material universes. In the material universe, God is present as the vibratory structure of all that materializes after agitations in one of the fields responsible for materializing particles, besides the movements of forces of nature. Each physical frequency of God represents a distinct divine quality and different agitation that produces and influences the smallest particles and emotions, as well as galaxies and linear time.

The diversity of divine qualities is expressed through various forms of worship across different cultures. These expressions take the form of triads, not just as countless gods, that represent the divine, the spiritual, and the material. For example, in the Hindu pantheon, Brahma, Vishnu, and Shiva are revered, while ancient Egyptian cults held the triad of Osiris, Isis, and Horus in high esteem. Various pagan trinity goddesses are found in European cultures, and Buddhism has its own triad of bodhisattvas. Similarly, Christianity venerates the Father, Son, and Holy Spirit. These deities embody different aspects of divinity and influence all aspects of creation, representing the divine, the spiritual, and the material aspects of God.

From a didactic standpoint, it can be said that the concepts of physical strength, determination, conquest, and impulse are often associated with God's longest oscillation. Many cultures have allegorically formulated gods of war to represent this idea, such as Ares and Mars in Greek and Roman mythologies, Hanuman in Hinduism, Montu in Egyptian mythology, Tyr in Scandinavian mythology, Huitzilopochtli in Aztec mythology, Ogum in Yoruba mythology, Hachiman in Shintoism, and Camalus in Celtic mythology. These deities are often regarded as embodiments of this first frequency. These deities represent comparable divine qualities, and thus, the gods of war serve as metaphors for the physical strength and agility bestowed by this divine frequency. Similarly, 'gods of thunder' or 'goddesses of beauty' were allegories created to elucidate other divine characteristics. Thunder, in this context, represents a phenomenon that paralyzes those who hear it and symbolizes strength and energy bolder than most other manifested events. As such, a god who is the ruler of thunder represents a facet of God related to justice, power, and kingdoms. On the other hand, goddesses of beauty personify divine qualities related to harmony, grace, and self-worth. Beauty is a concept that drives devotees to appreciate and cherish. In ancient times, before the advent of written language, cultures spread their religious beliefs through mythology, parables, and illustrations. For instance, when mythology depicts a god or goddess of war holding a sword, patrolling a specific location, and perhaps wearing red clothing while defying and overcoming everyone and everything, the worshiper idealizes that deity. They naturally resonate with those divine qualities through their thoughts, emotions, and behaviors, thus embodying those characteristics and becoming closer to or reconnecting with that divine frequency. This was the intended purpose of most religions—to help individuals reconnect with God.

Solar Logos and Planetary Christ

The solar logos, or archangel of a solar system, represents the highest command of a constellation that intimately influences all spiritual and physical activities within the system, including each planet and satellite. As the highest authority below God, archangels hold the responsibility of overseeing the operations and expansion of galaxies, condensing divine vibrations into both spiritual and physical realms, forming stars, and distributing energy, forces, and matter throughout the system. They project, create, and coordinate all aspects of creation in a solar system, and without them, the physical universe would cease to exist. Moreover, the key role of a solar logos is to create and maintain balance within a solar system. Although solar logos exert an incredible influence, they have no distinct form, name, or size, making it challenging to differentiate them from God. As the highest known form of evolution, solar logoi (plural of logos) continually evolve indefinitely and will do so perpetually.

The solar logos will also anoint planetary logoi, which are the archangels and therefore governors of a planet. As the condenser of the solar logos' emanations, the planetary logos breathes life into the planet, commanding its formation and maintenance in both the spiritual and physical realms. Moreover, the planetary logos determines the planetary cycles, including those of different humanities and geologic and atmospheric phenomena, and facilitates the transmission of divine light and love to all beings on the orb. In simpler terms, planetary logos are archangelic condensers of divine cosmic energy, rendering it compatible with the spiritual and physical needs of all beings on the planet. Life, in one form or another, is present on all planets, whether or not they harbor simple or complex physical life, such as Venus, Jupiter, Saturn, and beyond. It is important to realize that life goes beyond the limits of the physical universe and the limits of the third dimension. Therefore, life exists on all planets, albeit on varying vibratory planes.

The current planetary logos of Earth has another entity as its direct representative in this work, known as Sanat Kumara. Below Sanat Kumara is the entity known as Maitreya, the current Planetary Christ. The 'Christ' is a title, not an entity in itself, and Maitreya serves as the custodian of Christ consciousness. Jesus of Nazareth, who incarnated[1] as the Planetary Christ to teach the path to enlightenment through his own example, took on this title for his mission on Earth, with Maitreya expressing Christ consciousness

---

[1] Even before his incarnation, Jesus was already a high authority in the spiritual sphere. It took millennia for his essence to condense into a form capable of inhabiting a physical body once again. During this process, numerous enlightened spirits reincarnated in the same era, expanding and illuminating the world's aura – something still perceptible to this day.

through him. Jesus' primary mission on Earth during this incarnation was to guide humanity on the path to the divine, serving as an equal rather than a figure to be worshipped. Thus, in fulfilling this role, Jesus was responsible for the spiritual awakening of the planet.

The current planetary logos governs all spiritual, geophysical, social, and even scientific activities on Earth, acting as the chief intermediary of the solar logos. Jesus, in turn, is a significant master of wisdom and one of the principal entities of the solar system, working in the fields of wisdom, harmony, love, healing, faith, and devotion under the guidance of Sanat Kumara (the representative of the Planetary Logos) and Maitreya (the current Christ).

The Buddha, whose last incarnation was as Siddhartha Gautama, holds a hierarchical position just below Sanat Kumara and above all other masters on Earth. Although he no longer plays an active role in the planet's evolution, he continues to spiritually guide all the other masters.

The evolution of these entities does not cease, and it is possible that Jesus will succeed Maitreya, who in turn will succeed Sanat Kumara as the channeller of the planetary logos.

While planetary logoi are considered the governors of a celestial body, a planetary Christ is a high ranking entity, like a planetary logos, who embodies the 'Christ' role as the world's teacher, usually reincarnating to spread wisdom. Devas and angels, on the other hand, are believed to play crucial roles in carrying out the logoi (plural of logos) commands in creation and guiding humanity. There are also spiritual entities who incarnate for spiritual missions, so humanity can have more direct guidance and influence. Similarly, there are examples in which these spirits achieved enlightenment while incarnated. Although some may view this as a path to becoming an angel, it is likely that those spirits were already of a high ranking before his physical life.

Angels

In the universe, the hierarchy of spirits spans infinity. These spirits range from the most primitive forms of existence to those that have achieved high levels of enlightenment, transcending from an animalistic nature to that of angels. While there are angels that exist in various planes and have never been incarnated on Earth, this work focuses on elucidating the angels related to this solar system, as all universes are infinite.

The notion that humans have of angels differs from the idea that more evolved spirits have. Regardless of which religion first introduced the concept of angels and what this designation represents, it is important to note that angels were not made into angels, but rather, they were created as simple and primitive spirits. These spirits have experienced physical life on various orbs, ranging from group animals to individualized quadrupeds, humanoid beings, and even more sophisticated versions of humans on other planes of reality. As they learn and advance from their irrational, emotional, and selfish personalities, they substitute these traits with benevolence, altruism, and compassion. The more one is inclined towards love, the more angelic their status becomes. Angels possess more knowledge and freedom of self-will than ordinary humans and are continuously progressing, but despite the varying perspectives on who qualifies as an angel, they no longer require an astral body, as their mental body serves as their primary vehicle. Angels may fabricate appearances in case they want to appear to a clairvoyant, and they may often be similar to humans, although not necessarily limited to Earth's humans. Despite that, they typically possess a refined physiognomy and a shimmering aura. Not surprisingly, they often do not portray a specific form but rather only manifest as a sphere or bolt of light.

While angels reside in a permanent state of bliss, they

are beings that strive to support others in achieving higher levels of enlightenment, as mercy and fraternity are their ultimate purposes. Although angels are compassionate beings dedicated to helping others, they also have their own realities and personal goals apart from their functions of love.

Angelic hierarchy spans across several categories, ranging from enlightened beings liberated from the necessity of reincarnating or being tested to those who undertake significant roles on the planet and in the solar system. Some angels have even taken on physical incarnations as eminent avatars, playing crucial roles in human religiousness. For further information about angels, Jinarajadasa's (1921) work is recommended.

Angels do not become angels through any form of ceremony, just as wise individuals do not attain such a status through coronations or exams. As spirits, they continue to grow and may find themselves naturally belonging to a hierarchical group of similar beings. While angels do not refer to themselves as "angels," they are aware that humans would classify them as such beings. In this regard, the known Christian angels could certainly be actual angels, not because humans created or named them but because those benevolent spirits decided to assist those who worshiped the angels of religious literature. It is important to note that while humans have religion, spirits, and thus angels, do not.

Spirit Guides

Spirit guides are an important aspect of the spiritual enlightenment of individuals and collectivities and are, typically, the spirits of humans who are in the process of developing their abilities to become future angels. Essentially, they act as mentors to incarnate humans, typically those who have been their relatives, friends, or collaborators in previous lives or during their discarnate states. The spirit guide is thought to have a similar level of enlightenment as their protege, although they may possess more knowledge and training for the task at hand. While all humans are believed to have a spirit guide, it is important to note that this does not necessarily mean that this particular spirit is only assigned to one person. In most cases, a spirit guide has other duties as a trainee, and they may mentor other familial spirits as well. Additionally, animals are also thought to have spirit guides, but these guides are responsible for large groups of animals rather than one-on-one guidance. For them, these spirit guides are typically occupied with their discarnation and reincarnation processes, although they may also influence groups of animals to evolve towards intellect and emotions.

Spirit guides play a crucial role in programming an individual's future incarnation, more as a friend than a judge or teacher. Their help includes communication with the person's future parents, the programming of one's genes, the choice of their primary tests and lessons in life, and ideas for their discarnation. A spirit guide can aid an individual in many ways, but their task and power are limited to helping the incarnate human progress spiritually by encouraging better choices and the decline of harmful habits, thereby helping the individual be strong in the face of their karmic tests. While a spirit guide may not always be physically present with the person, they are believed to be aware of most episodes that happen to them, as they can see their thoughts and what is imprinted on their aura, in addition to being connected to one another via light threads.

Spirit guides can often be seen or heard by those who are mediums or psychics. Dreams may also show them; however, the physical brain will typically associate them and their familiarity with someone else that the person knows in the flesh.

Upon discarnation, it is believed that the individual may first encounter their spirit guide, although they may not immediately remember how familiar they are. As humans have their own spirit guides, it is also true that large populations, as well as the planet, solar system, and galaxies, have their own spirit guides as well.

Devas and Elementals

The words used in esoteric and spiritual literature, including in this book, to describe beings that go beyond what a simple human mind can understand, are subject to debate. Terms such as gods, devas, archangels, and angels may imply that a being must belong to a certain category before it can belong to another. The terminologies used to describe all beings must be understood as human constructs rather than these entities actual titles. Overall, the best way to categorize these beings, if at all, is through the idea of sublime hierarchies. This idea does not imply that one being is superior to or the commander of others, but rather that they have collected merit through consciousness expansion, work, and selflessness.

The concept of a god can refer to either the creator or the ruler of something, or both. Along these lines, devas can also be considered gods, not because they descended from God as minor gods but because they sustain and regenerate creation. The Sanskrit term deva translates to "shiny beings" and is commonly used to depict radiant entities that bear similarities to angels and archangels. However, devas possess several distinguishing characteristics.

In terms of their evolution, they do not reincarnate, irrespective of the plane of existence they inhabit. Instead, their spiritual growth and enlightenment occur over time through their own endeavors. Unlike angels, who are often traced to a humanoid past, either on Earth or elsewhere, devas are believed to have originated from elementals, although this cannot be considered a universal principle. These subtle beings are often described as light ballasts of about 15 meters (50 feet) or more, but they can also take on a humanoid silhouette or that of an entity belonging to a particular religion if necessary. Despite this, they rarely reveal themselves to humans.

Due to their elusive nature, literature about the devas is scarce and often infused with mysticism. They are incredibly subtle beings, which makes it challenging for one to study and understand them fully. Nevertheless, the devas are primarily responsible for the ministries of nature and serve as the spiritual vitality of the planet's power points, such as oceans, mountains, forests, open fields, and rivers. They are cosmic portals of divine light, being the "supervisors" of single trees and educational and spiritualist temples to cities and large regions of the planet, depending on their hierarchies.

Devas cannot be channeled by mediums, and it is not possible to communicate with them telepathically either. However, they may convey information to someone if necessary through means that are difficult to describe but that result in an instantaneous, direct knowing. Their mode of communication is more advanced than telepathy and beyond human comprehension. Nevertheless, devas can inspire large groups of people or individuals with power and means to act for the common good. In terms of hierarchy, it is worth noting that devas can hold authority over other devas within the same field of nature. Meanwhile, elementals are typically subordinate to the commands of the devas. For example, if there is a deva spirit presiding over a forest or woodland, the local elementals would likely facilitate the transition between elements according to the deva's directives. Overall, devas are what can be understood of the evolution of elementals.

Once the elemental transcends its primitive nature, it gains the powers of the field it pertains to.

Elementals of nature are entities that reside within the etheric double of natural elements or the natural world, such as trees, earth, water, magma, and air. They serve as intermediaries between physical and non-physical energies that flow through nature and are directly connected to the subtle planes of Earth. Many of these entities are hybrid beings, appearing as a combination of a humanoid being and a plant or stone. Elementals are typically small in size, ranging from a few centimeters to 2 meters in height or width. Although elementals lack consciousness and can only communicate through commands, they are capable of perceiving intentions. As a result, they are of interest to witchcraft sorcerers who seek the powerful services that these beings can provide, despite neither understanding the consequences. Intriguingly, when ignored, these entities may intentionally disrupt one's environment.

As they evolve, they may become minor deva spirits and take on the role of rulers over specific areas of nature where non-physical forces condense into earthly forces. Depending on their class, some elementals can also incarnate as less complex animals in the physical realm. It is worth noting that discarnate animals, particularly marine animals and aves, can sometimes live among elementals before evolving to the point of migrating elsewhere in the cosmos or becoming minor devas. Many elementals ultimately become responsible for managing complex elements, larger natural objects, and animals.

By intentionally sending love to a natural spot in nature, a deva may manifest. For those who possess psychic abilities, they may be able to see these beings, while others will still benefit from their presence regardless of their ability to see them. Similarly, elementals have been known to assist individuals in finding employment and success, as anecdotal as it may seem. However, natural spots that have been degraded may also have the awareness of these beings. While they do not seek revenge, particularly the devas who are known for their pure compassion, it is possible that angels of cure, a type of deva, may not align with an individual who asks for their assistance in a time of need.

Despite the lack of popularity and understanding surrounding these subjects, it is important to acknowledge that non-physical beings such as devas and elementals are indeed real. While they cannot be easily perceived by humans or by physical eyes, it is important to recognize that life extends beyond the physical and visible realm.

Karma Agents

Spirits whose task is to manage the karma of an individual or a group of individuals are known as "karma agents." Karma is a universal law, which means that it is as natural as the universe itself. Karma, also known as the law of cause and effect, was not created separately from what exists, and it applies to most beings, particularly those who are to experience lessons and tests based on what they need to learn or have caused. Although the law of cause and effect is a natural process in the spiritual universe, one's karmic lessons and expiation may be attenuated or subdivided into several minor experiences, depending on their needs for spiritual progress. Karma is neither a punishment nor an inflicted pain that someone else obliges one to undergo; it is a set of experiences that one decides or agrees to go through for their spiritual purification. For example, if one abandons their son in poverty, it is likely that they will eventually regret having caused such pain and therefore may decide to experience a life where they do not have family themselves. Or, perhaps, they will agree to spend decades of their lives as the father or mother of a child with special needs, who may very likely be the one who was once abandoned. Thus, they emotionally compensate and allay the guilt for having lived a life without what was once thought to be a burden. Despite not necessarily feeling guilt during their lives, one may, upon discarnation, encounter all the mental fabrications of their lives following them. Without a physical brain, which may influence or interfere with feeling or regretting such actions, the auras, which depict memories, thoughts, and emotions, become part of their reality. Thus, guilt, shame, and a sense of having wasted a life of regrettable inertia are common to emerge, thereby leading them to seek karmic lessons for fixing their wrongdoings and lost opportunities.

The agents' actions depend exclusively on the merits and needs of their protege before they can act on how a karmic lesson or expiation unfolds. They implement lessons, including tragic or pernicious events, that direct individuals to purge their spiritual debts and impurities. For example, if an individual commits murder, their karma may not necessarily be their own assassination, but instead, they may experience a new life dedicated to their victim, such as a loving mother struggling to raise a disabled child. In such circumstances, dedicating one's life to a disabled child could be a form of karma where the former murderer has their own life deprived in order to dedicate their incarnation to their past victim. Along these lines, the karmic agent could facilitate the child's accident while also acting upon the child's own karmic lessons with a disability.

Karma agents permeate the astral planes by manipulating thought-forms and holographic probable events that one must experience. They may also hinder or facilitate the implementation of karma through energy manipulation, persuasion, and other means. These spirits are also known as guardians, but they differ from guardian angels, a concept mostly associated with Abrahamic religions, in that tutelary beings designated by God protect individuals or groups of people. Guardians are responsible for securing passages, temples, roads, and other interdimensional portals. Fundamentally, all religious temples have guardians who protect them from possible malevolent or senseless disembodied spirits. Roads, squares, cemeteries, and natural spots such as beaches, woods, and waterfalls also have guardians when these places are located in areas with a considerable flow of people. If these places are found in the wilderness or have no consistent human traffic, they are only endowed with their own nature devas and spirits of nature, both of whom have not lived as humans, unlike the karma agents and guardians mentioned here.

These spirits play various roles, acting as karma agents, guardians, and spiritual workers in different religions and cults. They assume diverse names and appearances based on the culture and beliefs of those who invoke them. In mediumistic cults of Asia, they may appear as monks or yoga masters, while in American or Brazilian folk religion, they may take the form of barons dressed in suits and capes and holding cigars. In French séances, they may appear as local doctors or professors, depending on the form they have taken in their past lives. Depending on their field of work, these spirits may appear in various forms, such as an all-powerful king, a seductive dancer, or even a skeleton. These distinct appearances are intended to convey their defining characteristics. For instance, a skeleton is often associated with graveyards, where they serve as protectors, cleaners of nefarious energies, and guides for the spirits still bound to their decaying bodies. As skeletons, they may also be reminders that they once inhabited a physical body that they saw decomposing without being able to leave. Conversely, the sensual dancer represents a spirit of freedom, joy, and a lust for life. Typically, these spirits may have lived a life as a courtesan in difficult circumstances but now seek to help those struggling with low self-esteem develop self-love and the will to live. All of these manifestations represent their astral plasticity, conveying messages to those who still dwell in fear or oblivion. As the closest spiritual beings to incarnated humans, they are much closer to the general public and may even retain some mundane habits. Essentially, they are the spirits of individuals who may have led regrettable physical lives as criminals, arrogant politicians, pimps, or hustlers. Once repenting in the spiritual world, they choose to dedicate themselves to those who may be on the course of making the same mistakes they made in the physical world. As workers for the divine, they learn and gain merits for their own personal purification. By working on various fronts, they dismantle low magic spells, cleanse degraded areas, and promote vitality and the desire for life.

These entities typically undergo training to learn how to manipulate energies and conduct themselves with strict discipline, under the guidance of higher-ranking guardians in their hierarchy. Furthermore, they often align themselves with a higher deity, working under their vibratory umbrella. As a result, these guardians are usually the ones who collect the etheric double from offerings made to the deities they serve. As a karmic agent, this spirit may not have the power to alter one's lessons, but they do have the capacity and authority to influence how karmic events are experienced. In another example, a person who deforests an entire region out of utter greed could, presumably, experience destitution in their next life. However, if the individual progresses spiritually and helps nature recover, their future hardships may be mitigated, and the karmic agent may promote a prosperous event in their life. Although the individual's karmic imbalance may be reduced, the karma agent may also consolidate the lessons into a single challenge, which allows the individual to be released from karmic lessons sooner. On the other hand, if the individual is meant to accomplish a specific purpose in that lifetime, the agents may delay any adverse consequences until later in life, as a possible episode of poverty could hinder their achievement. Karma, or the law of cause and effect, can indeed be modified and reduced if the individual initiates benevolent deeds such as charity, forgiveness, and acts of compassion.

Karmic agents cannot be bought or persuaded. Instead, they take into account one's merits, debts, and spirit guides' opinions before implementing any type of "positive" or "negative" event in one's life. Karma is typically a lifelong condition or situation, and the karma agent ensures that such conditions persist. Sometimes, an incarnated individual may fervently pray for specific improvements in their situation, but their spirit, detached from the physical body during sleep, understands that the karma is necessary. As a result, they may contradict their unaware physical persona by asking the heavens to maintain the karmic lesson. In either case, the karmic agent is to follow the orders from "above", where, as mentioned, they depend on one's merits and spiritual needs before acting on them. Similarly, these spirits ensure that positive situations unfold in one's life. Lottery winnings and objects that are dropped and prevent individuals from encountering accidents are often orchestrated by karmic agents, typically by means of mental suggestions or influence on one's voluntary movements.

In conclusion, as the primary intermediaries between individuals and their karmic lessons, karma agents directly participate and act on the intricate workings one's incarnation.

Spiritual Vampires

Individuals who have lived a life consumed by addictions such as smoking, drinking, and excessive meat consumption may continue to be dependent on these habits in the astral planes. While some individuals may expel these dense fluids from their spiritual system, similar to how a diseased person may excrete an illness, others may continue to seek these sensations even after finding themselves in the lower zones of the astral plane. As a result, they may remain close to the astral counterpart of the planet Earth, inhabiting it not in its physical form but in its etheric double, in search of substances they only had in their physical body. However, once these spirits find the substances, they have no means of smoking, drinking, or eating them in their astral form, and therefore these substances do not produce physical sensations. Instead, the discarnate and addicted spirit can approach incarnate individuals who have the same addiction. Through their semi-physical aura, the sensations generated by the substances they take are externalized. This aura is composed of semi-physical elements, and as the physical body of those who consume drugs processes the physical molecules of whatever they consume, their aura exudes the by-products of those elements as semi-physical energy. In these terms, the discarnate spirit may parasitize the living human to obtain the sensation no longer readily available in the astral planes. Thus, the discarnate spirit will follow and incite the addiction for more of the sensations externalized in the aura of those who excessively consume drugs.

These spirits tend to occupy establishments such as brothels, bars, slaughterhouses, butcheries, skid rows, casinos, and fight arenas in the astral plane that is closest to these physical locations. These astral planes are superimposed and look exactly like the physical world. Spirits that parasitize other incarnate spirits can only do so if the incarnate spirit experiences sensations that correspond to the sensations sought by the spirits. This means that a person will only suffer from spiritual vampirism if the frequency of their thoughts, emotions, and sensations aligns with that of the addict's disembodied spirits. Pernicious habits, such as tantrums, rage, or envy, may also attract like-minded spirits who, in order to obtain vitality for their own desires to seek revenge against someone they know, will coax people to reproduce more of that behavior. These spirits absorb the torpid fumes from the auras of those they manipulate, assimilating the vitality in them because it may vibrate at the same rate as their own negative emotions. The more spirits absorb from people's auras, the more connected the two become, leading to a more intertwined dependence. However, the incarnate individual is not a victim but a producer of these negative energies and, therefore, is responsible for their own energetic losses.

The spiritual vampires, parazites, or obsessors, as the designation implies, become fixated on the incarnate individuals they obsess for a variety of reasons, such as a desire for revenge or intense longing. They remain in close proximity to their counterparts, and in cases where they harbor hatred for enemies from the past, they may try to influence bad habits and exacerbate sabotaging thoughts already present in the victim's mind, leading to more significant problems for the obsessed. In instances of longing, the spirit may have known their incarnate counterpart from a previous life and, upon encountering them, sought to remain in close proximity despite the incarnate now inhabiting a different body. As the spirits operate on a different frequency and in a troubled mental state, they cannot differentiate between different physical bodies. These situations lead the obsessed person to have their energies absorbed by the obsessors, who do not do so deliberately but require this energy to maintain a presence by the victim's side.

Many of the obsessions one may experience are manifestations of their own past lives. When a person undergoes extreme trauma, loss, injustice, or severe illness in a previous life, the associated emotions may become crystallized in their spiritual body. This often manifests in the form of unconscious fear, self-sabotaging attitudes, and even physical conditions during their present life. The crystallization of past life experiences can unfold as a personality, appearing as though it is another person. In fact, all the personalities of previous lives can be understood to "merge" back into one's spirit. However, in cases where traumas, addictions, or conditions are still present, those personalities may become separated from the individual's spirit, appearing as though they are third-party beings. These sub-personalities remain connected to the incarnated individual, and, through their thoughts and emotions, they negatively influence the life of their main spirit. It is important to note that these personalities are not spirits; rather, they are crystallized emotions that appear in the form of a sub-personality that appears to be experiencing a looping of that particular situation, representing the individual who experienced the emotion. These sub-personalities are rarely aware that they are sub-personalities or that they are not a spirit but a fraction of it. Once these personalities are treated through means such as meditation, self-purification (such as forgiveness and charity), and other spiritual treatments, such as apometry, they can be reintegrated back into the main block of one's spirit.

In summary, energy vampirism and spiritual obsession are carried out by ordinary discarnate individuals who still yearn for physical pleasures or seek to influence the living. In many cases, spiritual practices such as meditation and charity may help remove these spirits from one's life. In other instances, these spirits may relinquish their harmful behavior and seek help from benevolent spirits who guide them to spiritual communities, where they receive treatment, education, and assistance in planning a new incarnation to make amends for past wrongdoings. Despite the fact that most spiritual vampires engage in harmful behavior, they are still deserving of compassion, and the desire to eliminate them should be accompanied by a wish for their recovery. It is important to note that those who experience spiritual obsession or vampirism are primarily responsible for their circumstances, either due to their current harmful habits or past actions that have caused harm to their current obsessors who seek justice.

Ghosts and Poltergeists

Ghosts are commonly believed to be the non-physical manifestations of deceased people, often referred to as their disembodied spirits. However, the term "ghost" is frequently misused to describe both the spirit of a deceased person and supernatural apparitions or objects that appear to be spooky or haunting.

Ghosts are thought-forms that are imbued with a condensed form of light. Their frequency is so low that they only find resonance with similar objects and thoughts, which means that only those who reside in low emotional and mental states are likely to experience an apparition. However, it is important to note that ghosts can only be perceived by individuals with paranormal abilities to see beyond the physical universe. In other words, one may be surrounded by ghosts without realizing it if they cannot see or feel them through psychic abilities. A ghost or spirit can also be perceived through senses other than sight, such as through a particular smell or sound. Although these smells and sounds are not physical, they are a part of the spirit's mental construction, or aura. For instance, if a spirit is closely associated with the scent of roses, they may carry that scent in their aura, allowing a psychic to perceive the spirit as rose-scented. However, this scent is not created by the physical molecules of rose fragrance but rather through energetic vibrations.

Ghosts are typically perceived as ecstatic apparitions or manifestations, and they do not possess the ability to think or act independently. If a ghost appears to speak or converse, it is likely that it is a spirit communicating through the ghostly manifestation. It is important to note that while spirits can create ghostly phenomena, they cannot be ghosts themselves unless they choose to appear as an inanimate, haunting object. Frightening spirits are often characterized by their apparent confusion, attachment to worldly concerns, and unrelenting distress. They may also display an eerie appearance, which can be a result of their mental disarray or their ability to manipulate their astral body. By contrast, other discarnate spirits may exhibit a range of emotions and experiences associated with their state of being without necessarily displaying frightening or disturbing traits. The human body is composed of multiple energy fields, including the physical body, etheric body, astral body, lower mental body, upper mental body, Buddhic body, and Atman. The etheric body is a semi-physical energy field that lies between the physical and astral bodies. The astral body is the original matrix of the physical body, meaning it serves as its mold in the spiritual planes. A ghost-looking spirit is, therefore, an astral body that may have chosen to manifest itself in such a way. The appearance of an astral body is highly malleable and can be modified through belief or mental techniques. The plasticity of the astral body allows spirits to present themselves in various forms, whether more familiar or unfamiliar to the observer. This means that elevated spirits can choose to appear differently to different people or in different contexts, while malicious spirits may intentionally adopt a frightening appearance to induce fear or gain power over their targets.

Other phenomena related to ghost-like presences include the creation of apparitions through the use of lower magic. Entities carry out these spells to create ghostly apparitions designed to terrify their victims by manipulating etheric fluids. The victims in these cases do not need to possess psychic abilities, as the apparitions are created from semi-physical particles. Entities often extract the ethers used in these spells from the victim's own etheric field, particularly if the victim naturally externalizes ectoplasm. This not only makes the spells more potent but also gives the resulting apparitions a more material-like quality. These apparitions are often lightless figures or silhouettes, since the amount of plasma required to create a solid form is immense and could cause harm to the victim. However, evil spirits do not have enough power, ability, or knowledge to extract such a large amount of plasma from the victim's etheric body, making it highly challenging for non-psychics to see these ghostly fabrications clearly.

The manipulation of these energies unfolds as they become more and more condensed, but they will not possess any inherent movement and will not endure on their own, meaning that they are short-lived and extremely rare. Although they are considered a waste of energy, as enormous amounts of ether are necessary for the fabrication of a single specter, they are deployed not only to scare but also to trigger traumatic memories in the victim. This can lead the victim to a loss of control or cause them to take drastic actions, such as moving out of the house to escape the apparitions. It is important to emphasize that these phenomena are rare and may only manifest if the victim has the ability to externalize such fluids. If the targeted individual sees the fabricated ghost, it is possible, though unlikely, that other individuals nearby may also witness the phenomenon since these objects are made of semi-physical materials. However, it is important to note that psychics and mediums do not necessarily need such activity to occur before they can perceive an entity. Their sensory organs act as extra-physical antennas, allowing them to assimilate and decode certain invisible events.

Overall, both incarnate and discarnate spirits can fabricate ghost-like objects for various purposes, including haunting. In slaughterhouses, for instance, the collective trauma and suffering of the animals being slaughtered can give rise to ghost-like energies. Similarly, in cemeteries, the sadness and grief of mourners can create a kind of energy that some people interpret as ghostly phenomena. Additionally, the deceased may release some of their etheric fluids in cemeteries, which can contribute to the production of ghost-like energies.

Ghost-like energies, or the phenomena usually associated with ghosts, can be understood as an episode where condensed light is manipulated by spirits with the aim to scare or harm the living. The experiences of those who witness ghostly presences, including unexplained noises and broken or moved objects, can be classified as poltergeists. Poltergeists are a extremely rare type of ghostly phenomenon that is associated with physical disturbances, such as objects moving or being thrown, doors slamming, and loud noises. In some cases, people living in a haunted house may also report experiencing physical sensations, such as pinching, biting, or tripping, which may be attributed to poltergeist activity. Typically, these spirits have been described as troubled souls that haunt specific individuals rather than a particular location.

It is through the use of ectoplasm, primarily that of children, that these enraged consciousnesses can materialize physical movements. Ectoplasm is a semi-physical, gaseous, and semi-gelatinous substance that mediums excrete during trance, deep focus, panic, or delirium. It is a combination of carbon dioxide, hydrogen, albumin, potassium, and glucose produced by the organic cells of the body. In contact with physical photons, it rapidly evaporates. In spiritualist hospitals, healers use this fluid, which the physical body produces and the astral body controls to some extent, as a vital source of energy to help materialize physical cells and tissues. Only a specific type of medium can externalize ectoplasm. Therefore, the poltergeist phenomenon is only present where individuals can excrete it, which the spirit can use to materialize movements. It is crucial to emphasize that not all mediums are aware of their abilities, leading to the assumption that spirits are solely responsible for the events. Children excrete more ectoplasm than adults. This ectoplasmic excretion is thought to be a result of their youthful vitality and the continuing reincarnation process they undergo, which is commonly complete by the age of seven. The light threads connecting their spirits to their new bodies remain loose during this stage of life, allowing for accidental excretion of ectoplasm without causing any loss of vitality or discomfort. Ectoplasm is invisible and can rarely be seen, except when observed as a translucent substance under low, red physical light. Although images of this fluid have been widely circulated, especially from the 19th century, many are photographic scams.

Spirits of poltergeists may be able to move the fluid around the house and project it at people and objects. They may materialize body parts or tools to displace the desired object, resulting in the categorization of the phenomenon of materialization. In the astral planes, spirits manipulate the victim's excreted ectoplasm, bridging the two planes to materialize objects such as shouts, slaps, and tossed objects. Most of the physical harm caused by them is, in fact, manifesting on the victim's subtle bodies, but that leads them to believe it occurred on their physical bodies. However, only individuals connected to these spirits, either through past lives or via the resonant thoughts and emotions of the children's parents, are affected.

Poltergeist events are exceedingly rare phenomena. Typically, physical forces can explain the exploding of light bulbs, the falling of objects, and other accidents that have been attributed to spirits. While there are accounts of genuine cases, many accounts of poltergeists are fraudulent or result from irrational beliefs that spirits are responsible for strange occurrences. Thus, it is highly recommended to approach the possibility of a poltergeist with thorough discretion.

In conclusion, ghosts are thought-forms or apparitions that are typically witnessed by individuals with the ability to perceive extraphysical phenomena, even if they are not aware of this ability. On the other hand, poltergeists will commonly be perceived by one's subtle bodies and rarely by their actual physical senses. Thus, there is no need to fear seeing ghosts, as individuals who possess this capacity may have been witnessing such phenomena since childhood. In other words, those who have never seen ghostly objects are unlikely to experience visions, even if visiting a cemetery after midnight or engaging in other similar mysterious activities.

PART 2

Mediumship and Psychic Perception

Mediumship and psychic perceptions are two distinct yet related experiences or senses that involve the exchange or becoming aware of information between different planes of reality. Mediumship refers to the ability to bridge communication between different planes, allowing for conversations between two or more individuals through the semi-physical and non-physical activities of the medium's aura. This means that a medium can be an incarnated human whose physical vehicle is operated by a spirit or a spirit itself that communicates with other beings from a higher plane. On the other hand, psychic perceptions or paranormal abilities refer to any form of extrasensory capability that an individual possesses without relying on the communication or presence of other spirits. In other words, a psychic does not depend on an outside entity, and their abilities are built into their soul, allowing them to independently perceive the realities of events occurring on another plane or energies at frequencies other than those in which they currently reside. Overall, both mediums and psychics depend on their energy vortices and auras to perceive extraphysical entities and other phenomena.

Numerous individuals who provide readings as if they were mediums are, in reality, psychics who perceive the information emanating from an individual's aura. By scanning the auras of their clients, which encompass various forms of the people they know, ideas, the scenes they have experienced, and sensory information, such as smells and sounds, the psychic may come to believe that they are being informed by a specific spirit. Although these psychics are not intentionally misleading their clients, they are not genuinely experiencing a mediumship event. Typically, these psychics will ask their clients if they know someone named "John," for instance, or if they have any information about a "blue jacket," as the surprised client confirms that their deceased father's name was John and that they had given them a blue jacket as a gift for their last birthday. The psychic may even convey a message from what they believe is John, such as telling the client that their father is proud of them. However, these are usually just the client's own ideas, memories, and thoughts, which the psychic has managed to perceive. If the client feels as though they need John's approval for something, or if they know that John is genuinely proud of them, the psychic will mention it and convey the message as if it came from John himself. Communicating with a deceased person is a rare and difficult occurrence, even for mediums. This is not due to any external constraints, such as guards or angels preventing them from doing so, but rather because those spirits are typically residing at such different frequencies that they are unable to connect with the living or even recognize that a reading is taking place, apart from possibly being far away. Additionally, deceased individuals are not entities. Although both a deceased person and an entity can be classified as spirits, entities have functions and thus may permeate the frequencies of reality more readily at will, while individuals who have lost their physical bodies do not. Furthermore, a genuine medium will typically act as an intermediary during conversations, meaning that they will channel the spirit and discuss the present and future in a detailed and complex manner, allowing the spirit to incorporate their own unique personality traits.

The Chakras and the Three Auras

The concept of chakras was first identified by early Hindu clairvoyants, who, after locating these more luminous wheels of energy near seven of the body's main glands, realized that the glands and the chakras had similar functions, albeit on different planes of reality. Although the chakra system was also observed by several other cultures, the Hindus were the first to mention them in their scriptures, which are well documented in the Vedas and the Upanishads; thereby, their nomenclature and school of thought have become the most common worldwide when discussing such energy portals.

The human body is composed of various glands that possess distinct subtle electromagnetic properties and serve different functions within the organism. Seven of these glands, including the pineal, pituitary, thyroid, thymus, pancreas, testicles/ovaries, and adrenal glands, are noted for being correlated to particularly strong subtle energy centers that direct vital force, chi, or prana currents to surrounding tissues and organs. Each center, also known as a chakra, reflects a unique electromagnetic wavelength and carries a unique function within the overall energy body of an individual. The top three glands, namely the pineal, pituitary, and thyroid, function as materialized centers where subtle energy and information traffic through. In essence, a chakra is a locale where energies are either condensed or made subtle; in other words, through the chakras, spiritual energy is condensed into a semi-physical form, while semi-physical energy is volatilized to the point of becoming spiritual fluid.

The chakras related to the top three main glands vibrate at a faster frequency than the other main chakras, and they operate in matters related to faith, intellect, thoughts, and spiritual purpose. The chakra associated with the thymus gland is located in the chest cage and is pertinent to the circulatory and respiratory systems, as well as relationships between the individual and others and one's morals. The thymus gland functions to absorb and externalize noble emotions, and thus the thymus chakra (also known as the heart chakra, as the two are in close proximity) governs emotions such as longing, gratitude, and appreciation.

The physical manifestation of the human body can be seen as an expression of the spirit, with each organ representing a specific faculty of consciousness. For instance, the eyes are the physical manifestation of the spirit's ability to see, and the ears represent the spirit's ability to hear. Similarly, the pineal gland represents the spirit's ability to communicate through thoughts and perceive different frequencies. According to the Philosophy Research Journal (1958, vols. 18–19), ancient cultures such as those in India, Egypt, and China have detailed references to the "third eye" in their sacred scriptures and stories. This term refers to the pituitary gland's extra-physical faculties, which are associated with the chakra that resides between the brows. It is a common mistake to confuse the pineal gland with the third eye, but they are not interchangeable. The third eye chakra is oriented forward, while the crown chakra points upward. While the pineal gland does bear a resemblance to a conventional eye, it should not be mistaken for the pituitary gland, which is associated with the third eye chakra.

The subtle, or non-physical, version of the pineal gland is sensitive to the number of photons in the electromagnetic field. This sensitivity causes it to produce a wave motion that ripples through the individual's entire biofield. Due to various scientific studies in evolutionary and comparative biology, neuroanatomy, and neurophysiology, the pineal gland has been identified as a dormant photoreceptor with a lens, cornea, and retina in its anatomy, similar in composition to that of an eye (Eakin, 1973). In addition to its physiological similarities to the eye, the pineal gland contains unique formations of calcite crystals, supporting ancient claims that the pine-shaped structure functions as a perception antenna. Based on a study conducted by Lang et al. (2002), the examination of twenty dissected human pineal glands revealed the presence of 100 to 300 calcite micro-crystals per cubic millimeter. These calcite micro-crystals are similar in composition to those found in the otoconia region of the ears, which transmit mechanical forces to sensory hair cells in the utricle and saccule. The calcite micro-crystals exhibit piezoelectric properties, which allow them to convert sound into electricity without requiring an external power source. According to Rosen et al. (1992), this piezoelectric effect is due to certain materials, like quartz, having a property that produces a continuous electrical charge when subjected to mechanical stress or, conversely, when undergoing mechanical deformation when subjected to an electric field. The pineal gland, also known as the epiphysis, plays a physiological role in differentiating light from darkness, hence its function as a melatonin producer. Melatonin production in the pineal gland is influenced by the presence or absence of light, with stimulation occurring in darkness and inhibition in the presence of light. However, Reiter (2016, 2019) indicates that the pineal gland serves as a backup source for melatonin regulation, as the majority of melatonin in the body is actually present in the mitochondria of cells and gets activated by sunlight.

 The pancreas, gonads, and adrenal glands are associated with the lower main chakras, which primarily deal with more tangible or physical concerns. The pancreas symbolizes the physical manifestation of the spirit's ability to interact with others and circumstances, as well as influence willpower. The gonads are responsible for the expression of physical creativity and physical pleasure, while the adrenal glands play a role in providing physical energy and reaction. Naturally, these glands serve as regulators of the body and play a crucial role in maintaining the functioning of adjacent organs. The main attribute of a chakra is that it is a communicator between the organs and bodily functions and the spirit and its settings. Such communication develops via energetic fields that can be physical, as in electromagnetism; semi-physical, as in etheric waves; or spiritual, as in light. Apart from developing physical repercussions on organs, tissues, and hormones, the chakras have a thorough influence on the psyche of the individual and are considered the touch point for mediumship and paranormal perception. Each type of mediumship or paranormal ability is carried out by the semi-physical counterpart of a specific gland; however, among the 7 main glands and their related chakras, the pineal gland appears to play the most important role in almost all categories of mediumship, whereas for psychic perceptions, the pituitary gland appears to play a major role, especially in cases of mental clarity and awareness. The physical glands are not necessary for the spirit, but only for the incarnated soul. In periods where the spirit is not incarnated, the spiritual glands, or para-glands, are present as a means to preserve the spirit's form and its nature during their evolution path as humans. Likewise, as some chakras develop different functions, others cease to exist, and others merely reside dormant.

As per the second edition of the Oxford Dictionary of English published in 2003, the term "vortex" is defined, in part, as a whirling motion or mass. As the chakras are considered wheels of energy, a whirling mass can naturally be observed around them. Since an uncountable number of chakras are present in both the physical and spiritual bodies, the quality and scattered mass of these vortexed fluids are what constitute what is commonly known as the aura. The human body and its vortices generate physical currents as well as semi-physical and non-physical fluids, which contribute to the formation of radiant emanations. These emanations are present in all beings across the spiritual, etheric, and physical universes, forming an integral part of the makeup of all living entities. Therefore, auras, radiant emanations, and biofield are common terms used to refer to these emanations.

Planet Earth has its own aura, also known as the magnetosphere, which is generated by electric fields, electric currents, and magnetic fields, or, in other words, charged particles. These charged particles include ions, electrons, and protons, which are primarily sourced from the solar wind and the planet's own ionosphere. According to Russell (1972), the interaction between these charged particles and the planet's magnetic field results in the formation of the magnetosphere, which serves to protect the planet from harmful solar radiation. Moreover, as asserted by Brandenburg (2007), the widely accepted dynamo theory posits that a celestial entity engenders a magnetic field through the rotational and convective motion of an electrically conducting fluid, a phenomenon that sustains the magnetic field over vast astronomical time frames. Compasses, for instance, function by utilizing the Earth's magnetic field, which causes the magnetized needle to align itself with the Earth's magnetic field and point towards the magnetic north and south poles.

The human body also generates radiant fields, which are thought to be analogous to the Earth's magnetic field and are produced by the movement of electricity within the body. Research by Korotkov et al. (2004) and Rubik et al. (2015) has suggested that auras are electromagnetic radiations that surround living beings and are composed of a magnetic component, with a person's physical aura representing their biofield.

Research conducted by multiple sources, such as Groner (2017), supports the fact that the human body produces substantial amounts of electricity. This electricity is generated through various means, including the electrical activity of cells, which occurs through the synchronized oscillation of microtubules, centrosomes, and chromatin fibers. The electric fields generated facilitate crucial biological events such as chromosome organization during cell division (Zhao and Zhan, 2012). The human body also utilizes electrical synapses, which enable communication between neurons (Pereda, 2014), as well as other cells that communicate via electrical signals, such as those involved in the heartbeat and muscle contraction. The electric currents produced by these processes generate a bioelectromagnetic field, also known as a physical aura or a biofield.

The human body also generates a semi-physical aura that is less noticeable than physical magnetic or electrical fields. This energy field is influenced by a person's thoughts and emotions and serves as the primary bridge between their physical and non-physical experiences. The chakras convert physical particles and waves into vitality and etheric energy, contributing to this aura. The radiation of thoughts also travels through the chakras, replicating the thought in one's aura. On the other hand, the astral aura is a fully spiritual radiation that reflects a person's mind and is composed of higher-frequency energies than the physical and etheric auras. The aspects of the astral aura are determined by the frequency of their thoughts and emotions, and it can interact with the auras of other incarnate or discarnate spirits as well as their non-physical environments.

The aura of humans who are currently incarnated can be thought of as three interwoven layers: the physical layer (the densest), the semi-physical layer, and the astral layer (the most subtle). These layers are created by the movement of particles and waves within each respective body. The astral body, which is a decoder of one's emotions, emanates a luminous and dynamic spiritual aura. Though imperceptible to physical eyes and instruments, it comprises the same components as physical light but pertains to a distinct plane of existence and frequency. The astral aura blends two types of light: astral light, which is denser in nature, and mental light, which is subtler in form. It serves as the initial point of contact between the medium and spirits during communication. Similarly, the etheric aura emanates from the movement of respiration and electricity within one's physical cells, coupled with the magnetism of one's thoughts. The etheric body functions as a bridge of energies between the astral and physical bodies, beyond its role as a glue for both. Cleansing and energizing herbal baths and incense smudging tend to act upon this body. The physical aura comprises a completely material emanation from one's atoms in the form of magnetism and light. It is vital to note that an incarnated human possesses six bodies, each residing on a different frequency. These bodies act as vehicles for one another. Thus, the astral body uses the physical body as a vehicle on the physical plane. The lower mental body employs the astral body as a vehicle in the astral planes, while the upper mental body uses these two bodies as vehicles. Since mental bodies exist as fields, they may not exhibit an aura. Hence, the mental body can be perceived as the emanations of an astral body. The buddhic body utilizes the upper mental body to manifest in the mental planes, while the atman, which refers to an individual's true portion detached from Source, resides in the buddhic body before merging back with Source following a lengthy journey. Among all the bodies, only the physical, etheric, astral, and lower mental bodies alter shape and, consequently, are replaceable. The other vehicles may never change. It is crucial to stress that the terms "body" and "vehicle" are merely utilized to convey the notion that an individual, though a unified whole, may be divided into various classes according to the various frequencies they reside at.

Mediumship

Mediumship is the ability to communicate with entities that exist on a different plane of existence. A medium acts as a bridge between the physical world and the non-physical realm, facilitating communication between spirits and individuals who are still alive. The majority of mediums are well aware of their circumstances; however, some find it difficult to fully comprehend their condition and thus unconsciously suffocate these events with drug abuse or persistent clinical treatments. Others, however, may develop mental illnesses for the lack of mental peace, not knowing that if they were educated about it and their mediumship was exercised, their lives and inner peace could unfold normally. Most humans have some slight degree of psychic perception, despite it not being obvious to them, but not as many individuals possess any degree of inherent mediumship, though it can be minimally enhanced with meditation. As is often assumed by those unfamiliar with their work, mediums are not limited to communicating only with the spirits of the deceased. They are capable of communicating with a variety of entities, including the non-incarnated and those who may never have been incarnated, from different realms, including those from other planets.

Mediums do not become mediums by chance. Any extrasensory faculty is an expression of the spirit onto the physical body; in most cases, mediums have a karmic debt or a mission, and the practice of free-of-charge mediumship will assist them in helping others and thus righting past wrongs. As such, if the spirit desires or is advised to develop such spiritual works during physical life, they may imprint mediumship characteristics in their genes before reincarnating. The relationship between spirituality and genetics has been a topic of interest among both the scientific and spiritual communities. Scientifically, Koenig et al. (2005) have suggested that religious tendencies and levels of faith may be reflected in an individual's genetic code and thus in their brains. Likewise, the presence of skepticism or irreligiousness can also be linked to genetics. From a spiritual perspective, the spirits choose to inhabit physical bodies that are compatible with their nature and future lessons, or are advised to do so. A spirit who must centralize their lives in their future incarnation away from doctrines and sects and must practice something other than religion will invariably reincarnate in a body that does not contradict such a traced path. Similarly, a spirit embarking on an esoteric or spiritual mission on Earth is unlikely to incarnate in a body that produces a skeptic's or materialist's brain. For karmic reasons, that is, for one's needs to experience a particular physical life, a specific family is chosen before the egg meets the sperm. This chosen family is typically one's spiritual acquaintances, and therefore they are likely to be part of what is called "group karma." A group's karma is a group of spirits who have had physical lives as family members, friends, spouses, and so on. They presumably share similar collective experiences, faults, and duties, and for these reasons, these spirits rejoin the same families to amend the unfinished lessons one has with another. Along these lines, the chosen parent's genes will commonly reveal a genetic code that has that very gene that is to be expressed or kept dormant in the life of the spirit to reincarnate. For these reasons, genes will invariably manifest the traits one wishes to manifest or repress. Thus, spirits are perfect matches for the physical bodies they inhabit. As mediumship and paranormal abilities have a propensity towards altered states of consciousness, it can be concluded that mediumship and psychic trance have both scientific and spiritual explanations, which complement one another and by no means contradict one another.

When the extra-physical faculties of the pineal gland are well developed, the most common type of mediumship would be that of mental communication with beings from other planes. Mental communication regards conversation via the means of blocks of thought, as opposed to the sound of speech or visual images. Such communication would commonly advance via the extra-physical functions of the pineal gland, which transform the non-physical waves of thought into semi-physical information. Subsequently, the chakra related to that gland would redirect the wave of information to its immediate neurons present in other parts of the brain. The decoding of the electromagnetic signals would then unfold in the regions of the brain responsible for creativity and imagination, as well as regions responsible for cognition and those that decode sensory electromagnetic signals, such as the auditory cortex and the visual cortex. Although the crown chakra is primarily responsible for this type of mediumship, other major vortices in the body, such as those in the third eye and throat chakra may also facilitate intercommunication and awareness.

The most notorious type of communication for mediums, at least in spiritualist temples and séance groups, is that of channeling. Spiritual channeling must be distinguished from spiritual possession. In channeling, the entity couples their aura with the medium's astral aura, thus forming a single field. In spiritual possession, the malignant entity supposedly links their para-chakras to the victim's, thus controlling their senses, including their synaptic activity, and compromising their behavior and emotions. In most possession occurrences, the spirit is so united with the victim that abruptly removing them from the victim's aura would cause them to physically die. Furthermore, in spiritual channeling, the spirit does not enter someone's body but instead stays beside the medium at all times.

To channel a spirit or to come under its channeled influence, several methods can be employed. For instance, the mediums in a particular group may chant or sing in a predetermined rhythm, inducing their brains to adapt to the sound patterns and reside in waves typically associated with altered states. Similarly, shaman drumming and other animistic religions utilize sounds that are paced to shift the brain waves from beta (indicative of a wakeful state) to approximately 5 hertz, which denotes theta waves or the dreaming wave. Altered states induced through various means can lead to an expansion of the auric field, which is necessary for spirits to draw near and be channeled. In the same vein, music can serve as a tool to aid focus both before and during the process of channeling. It not only induces a shift in brain waves but also helps to maintain the individual's attention on the character of the music, often aligning with a spirit that resonates with the theme of the music.

The densest aura field of a spirit is identical to the astral aura of an incarnated human. If the auras of both individuals come into contact by mutual agreement, they can form a shared aura, allowing one to influence the other. The thoughts and commands present in the aura field of the spirit are then absorbed by the astral aura of the medium, perceived by their semi-physical aura, and finally captured by their chakras. The chakras function as energy transformers, converting the semi-physical information into electrical signals for the neurons. As a result, the information contained in the aura of the spirit is transferred to the aura of the medium and condensed into electrical signals, which ultimately influence the medium's behavior. For a full channeling, the energy vortices of the medium must be interconnected with the spirit's, so the spirit can control the body parts they need in order to externalize some form of communication. The connection of the medium's chakras to the spirit's is done by magnetism, whereby they magnetically pair. It is rare to observe energy threads like the ones seen when the astral body separates from the physical body during astral travel or sleep. When most chakras interconnect, the spirit gains direct access to the body regions that that chakra is responsible for. When the spirits couple their auras, but only their top three chakras, with the medium's, they will be able to access the medium's mind and speech organs, but they will not have control over their bodies, which means that the medium might not acquire personal traits, such as a different walking style or an ability, like painting, for example. In circumstances where the spirit couples their respective auras but is only able to pair their lower chakras with the medium's (due to the medium's lack of concentration), the spirit is able to influence the medium and their moves and manners, but comprehensive ideas and thoughts may not be completely understood. In other words, the medium's actions and activities will not be controlled but merely influenced by a third party. In channeling, the medium will always have total control over what occurs, as the spirit does not and cannot impose their power on the medium's will. As such, all of the channeling is performed by the medium, whereas the spirit simply suggests ideas, speech, and movement to them. Ultimately, mediums listen to and feel the entity and thus replicate their words and behavior.

Other types of mediumship events include psychography, also known as "spirit writing" or "automatic writing," which is the faculty of allowing a spirit to use one's arms and hand to write. In psychography, the medium is aware of what is being written; however, they have little power to prevent a word from being written or to add something that was not intended by the spirit writer. The spirit writer will typically couple their aura with the medium's, which serves as a foundation for the two to interact and better perceive one another. The spirit then links their heart chakra to the medium's, as well as their scapula chakra, a minor chakra located on both scapula bones (known as shoulder blades). On both scapulae, these chakras develop the tasks of mediumship on the left and psychic perception on the right. When any mediumship phenomenon occurs, the left-hand chakra is normally connected to the spirit via magnetism. If the medium has the right-hand scapula chakra more activated than their right-hand one, they may imprint a more personal trait on the mediumistic event, such as old-fashioned handwriting in psychography, a different tone and accent in psychophony, or even a walking style in full channeling, revealing an animist style.

In a similar vein, psychophony is the ability to allow a

spirit to use one's speech organs to communicate. This form of mediumship is often mistaken for spiritual channeling; however, channeling only occurs if there is an aural coupling and all the seven main chakras of both parties are connected, whereby the spirit has full control over the medium's physical body. During psychophonic activity, the spirit connects their throat chakra to the medium's and, often, other chakras if the situation requires. In psychophony, the medium's control over their own speech is lessened, meaning that the spirit may use it to vocalize whatever it pleases; nevertheless, the medium can still refuse to speak particular words if they decide not to. The replication is based on word-for-word reproduction rather than tone or accent similarity. Present in most types of mediumship events, animism.[2] It typically occurs when the medium perceives the spirit's compartmental traits and attempts to replicate them, often exaggeratedly. Along these lines, in psychophony, the degree of animism is typically much lower than that found in spiritual channeling. Both psychography and psychophony work by converting non-physical information into semi-physical electrical signals, which are then converted into mechanical movement.

Intuitive channeling is the most common type of mediumship, in which a spirit facilitates the transmission of telepathic information while the medium articulates, writes, or conceives ideas as if they were their own. These mediums possess a seemingly innate awareness of information with no recollection of its origin. This contrasts with "spiritual influence," which emerges from suggestions. In intuitive channeling, the medium discerns that the information is not their own and has originated from an external source. Suggestions, on the other hand, comprise

---

[2] - Animism (from Latin, animus, "soul") is a set of mental phenomena or behaviors produced by the person themselves, without external interference.

phrases directed at the recipient, which may be assimilated as their own depending on how well they align with their ideas. In intuitive channeling, the medium replicates most of the entity's communication, not verbatim but as a complex collection of ideas. This type of mediumship is infrequently discussed as, in numerous cases, it cannot be distinguished from the medium's own ideas and does not exhibit any demonstrable phenomena that attract curiosity. Intuitive channeling appears to be prominent in those who meditate on a regular basis.

Psychic Perception

Psychic abilities refer to the ability to perceive and interpret extraphysical elements such as entities, impressions, and energetic traces. An individual who experiences psychic perceptions or has the ability to see, hear, or feel something that is attributed to other planes of reality is someone who is likely born with the condition, even though it can also be developed and enhanced through exercises and meditation. When the spirit incarnates in a physical body, their abilities to perceive other planes of reality are typically reduced. Inhabiting a physical body means being able to reside in the physical universe, and as such, one is not supposed to consciously dwell in and experience multiple realities simultaneously. Psychics, however, often experience fragments from another plane. They may see, hear, smell, feel, and think according to the information from another frequency. Like mediums, psychics cannot choose when these extra-physical phenomena occur, but they can train themselves to avoid certain experiences and enhance or broaden others.

The mechanism that distinguishes a psychic from an ordinary person is that the astral and etheric bodies of psychics are more easily perceived by the physical bodies they inhabit. Their nervous cells and glands are more attuned to the subtle waves of the other planes, where the astral and etheric bodies are natural. The neurons of an individual who is known to be psychic are familiar with subtle forms of what can be understood as electricity, whereas the neurons of those without any extra-physical capability interpret these waves of information as radio frequencies that only give off information if one is actively tuned into them. For the cells of ordinary individuals, these emanations are too subtle and almost impossible to be perceived and assimilated into electrical and chemical operations.

Telepathy, although not as popular as clairvoyance, is in fact the most common variety of paranormality. Telepathy is a common mode of communication between spirits who exist on the same frequency or plane of reality. It involves the transmission of information from one mind to another using extrasensory perception. When the extraphysical field of an individual absorbs a wave of information from another mind, it is transformed into electrical and mechanical activity in the brain of the receiver, provided they are incarnated. In cases where the communicators are incarnated, a block of thought externalized by one is projected and absorbed by the counterpart, who assimilates the information as their own. The block of thought, or compressed ideas, is projected onto the emitter's aura. As the mental sphere of influence of the telepathy receiver is less affected by being immersed in a physical body, they can unconsciously scan the emitter's aura and promptly absorb the block of thoughts. Therefore, a large set of data can be absorbed and understood more easily and quickly than in a typical conversation. An entire story can be sent telepathically in a fraction of a nanosecond and comprehended without the need for reflection or the use of formal vocabulary. The receiver obtains integral information, containing all kinds of details such as context, images, sounds, scents, emotions, and timeline. Telepathic phenomena can manifest when an individual is capable of perceiving the aura of another person and erroneously believes that they have accessed their counterpart's minds. Within the aura, thoughts, ideas, and memories may manifest, and individuals endowed with the ability to perceive other people's semi-physical or spiritual auras may be able to discern the fluctuating objects. It is noteworthy that telepathy is most frequently observed between individuals inhabiting the same spiritual plane, and it constitutes a prevalent mode of communication on spiritual planes.

Clairvoyance is another psychic ability that is frequently used in divination. It involves the paranormal capability of visually perceiving objects, entities, and locations from other planes of reality. It allows individuals to "see" beyond what is physically present, offering insight into unseen realms of existence. The term "clairvoyance," which originates from the French terms "clair," meaning "clear," and "voyance," meaning "to see," pertains to the present moment. Retrocognition, often known as "memory regression," is the perception of images from an individual's own past; this might occur in reference to either the past of this life or the past of prior lifetimes. If the pictures regard the future (as supposed, likely, or relating to it), the phenomenon is called precognition, also known as premonition. What the physical eyes see pertains to the physical plane as a result of the light that reflects on objects. Differently, what is seen in clairvoyance phenomena pertains to another plane that is not dependent on a luminous source in the physical dimension. Although the clairvoyant may use their physical eyes to "see" that which pertains to another dimension, such phenomena are a result of being in the physical body and thus having physical eyes. The physical conditioning of the eyes may lead the clairvoyant to feel as though their physical eyes are actually witnessing such apparitions or scenes; nevertheless, it is their aura fields that capture such images from another plane. Furthermore, one can clearly see something from another plane while keeping their physical eyes closed. Simply put, the aura fields capture the images, which are subsequently passed to the auric fields of the pituitary gland, which then sends the images to the rest of the brain, including the occipital lobe—an area responsible for vision and eyesight.

Similar to clairvoyance, clairaudience is a paranormal ability to perceive sounds from other planes of reality or, on occasion, to decipher information received through auditory channels. Although these sounds are perceived as an internal mental activity, they are often described as a loud voice inside one's head. The sounds are not transmitted by physical waves, but the individual receives mental phrases from another entity, which are then assimilated in the parts of the spiritual brain related to hearing and subsequently integrated into the physical brain. Empirically, clairvoyance is associated with the "third eye" chakra, while clairaudience is linked to the throat chakra as its primary receptor. However, both the crown and third eye chakras also play important roles in the development of these abilities.

Psychometry is a phenomenon that may be less well-known but is equally interesting compared to other types of psychic abilities. It involves perceiving the mental emanations that are attached to objects as well as spaces like houses, which correspond to the past scenes of individuals or the history of an item, location, or scenario. Episodes of psychometry may occur deliberately or not, whereby the individual feels and visualizes who previously owned or significantly magnetized the object. In the scheme of an aura, mental propagations of someone's thoughts are imprinted on objects, just as scent from a soap bar is released in its environment or onto the objects nearby. When the paranormal touches the objects, they feel the residue of radiation as though they are smelling their scent. As their aura senses the magnetic signature of the thoughts imprinted on the object, they may easily connect to the owner of such radiation, that is, to the individual or individuals who emanated that radiation. The objects or their vicinity are not connected to the occult individuals, as in a voodoo-like connection, nor are they intertwined perpetually, but a fragment of their energy remains attached to other physical objects. If the object belonged to different people, the more difficult it will be for psychometry to unfold clearly, depending on the psychic's intuition for precise answers.

When examining the topic of mental propagations, future occurrences may also be perceived through the emanations that the holographic forms release. When the thoughts of the mass or those of an individual produce an object with enough magnetism, it may manifest in the more ethereal realms before it materializes in the physical plane. Along these lines, episodes of precognition may arise in those with the ability to perceive probable future occurrences. Precognition is the precise term that explains the foreknowledge of an event, where the psychic has the ability to see a future occurrence. Premonition, however, is a strong feeling that something unpleasant might occur. Precognition is closely linked to the extraphysical capacities of the pineal gland, pituitary gland, and pancreas and their relative energy vortices. Similar to foreseeing and clairvoyance, precognition is often experienced in dreams, particularly in cases that involve a large number of people in a tragedy. In the same way as clairvoyance, the individual captures the probable future scenes as they are planned according to the laws of cause and effect. Pemonition is often mistaken for episodes of déjà vu, which refers to a perceptual phenomenon characterized by the subjective impression of having previously experienced a particular event or situation. Déjà vu is believed to result from a functional alteration in communication between the amygdala and hippocampus, two structures within the brain that process emotional memory (Richter-Levin, 2000). The hippocampus may receive events in the sequence before the amygdala has completed processing the previous episode, resulting in the sensation of having already witnessed that scene. From a neuroscientific viewpoint, the conscious and unconscious minds are differentiated based on the regions of the brain associated with them. The conscious mind would be related to the cerebral cortex, which comprises seventy billion neurons, while the unconscious mind would be linked to the cerebellum, which has thirty billion neurons. The conscious mind focuses on the most significant aspects of awareness, while the unconscious mind operates through intuitive or parallel logic, including repressed feelings, automatic skills, subliminal perceptions, and complex phobias and desires (Sternberg, 1999). Conversely, the spiritual perspective suggests that the human mind has both physical and non-physical aspects. The physical mind encompasses self-identity and feelings, while the non-physical mind is believed to contain eternal memories and individual identity beyond the current incarnation. The conscious and unconscious minds are thought to interact through the subconscious mind. The conscious mind separates time from space and linearizes events, while the unconscious mind perceives time and space together, or cyclically. Déjà vu may be seen as a discrepancy in communication between the conscious and unconscious minds; however, déjà vu also seems to be the least paranormal activity among other phenomena, as premonition has few to no similarities with déjà vu. While most instances of precognition do not materialize, a psychic is usually able to determine whether the images are coherent enough to manifest in physical reality. On the other hand, accurately perceiving future possibilities is heavily reliant on intuition, which is less likely to be inaccurate.

Intuition may or may not be described as a psychic phenomenon, depending on how it is described, for it is the most physical of the abilities. Instinct, as described by Charles Darwin in "The Expression of the Emotions in Man and Animals" (1872), refers to a spontaneous reaction demonstrated in animal behavior. It is characterized by an inner impulse that prompts a living being to act without self-awareness. And, naturally, there is a difference between instinct and reasoning, which are fundamental aspects of human behavior and thinking. Instincts are primarily focused on self-survival or the survival of an offspring and are linked to physical, genetic, and environmental factors. Reasoning, as defined by Piaget (1967), involves the use of logic and critical thinking to draw conclusions from new or existing information. It is a cognitive process concerned with decision-making, deduction, and the understanding of sensory information or abstract concepts. The branch of philosophy known as logic studies the ways in which humans can use formal reasoning to produce valid arguments. In these terms, intuition can be considered a complex combination of both instinct and reasoning, as it involves prior knowledge and sensory perception. It is perceived through the third eye chakra, which serves as a "seer" of reality by perceiving the subtle aspects of experiences beyond the veil of illusion. The ajna chakra, from Sanskrit "center of command," plays a crucial role in transmitting this information through the aura and into one's semi-physical aura, where it can be subsequently decoded by the brain. As such, the clarity of intuitive information can be influenced by a person's pre-existing beliefs and illusions. For example, those who focus solely on material success or societal status may experience weaker intuition. A fleeting glimpse of reality and a feeling akin to a forgotten dream are what intuition is like.

Conversely, individuals who have a more expansive and open view of the world tend to receive clearer intuitive signals.

The third eye region allows individuals to perceive reality as it is, while the solar plexus chakra enables them to feel the qualities of that reality. The navel chakra also plays a crucial role in perceiving the unseen, but more in an energetic sense, helping individuals identify semi-physical energies and their aspects. However, it may not be the best antenna for detecting exclusively spiritual frequencies. Notwithstanding, when the navel chakra is activated, it may result in the manifestation of telekinetic abilities.

Telekinesis is defined as the ability to materialize subtle energy and manipulate physical and semi-physical elements using the power of the mind. This rare paranormal capacity is usually revealed in subtle manifestations, where the psychic is only able to move small, lightweight objects such as a thin leaf or a piece of paper. Telekinetic abilities are a result of the emission of currents from the navel and heart regions, which extend to the object's radiant field, ultimately promoting movement through an intensified magnetic force based on the individual's intention. It is important to note that the ability to manipulate objects with the power of the mind is limited to close proximity, and any movement at a greater distance would be considered a quantum entanglement event rather than a psychic phenomenon. Overuse of this capability may pose a risk to one's heart health and result in premature death. This occurs when the heart chakra is compelled to donate significant amounts of etheric fluid to one's solar plexus region, as the energies from the solar plexus chakra are externalized to produce physical motion. condense and produce physical forces that move an object. Although the solar plexus chakra, known as manipura, provides most of the forces used in materialization, the heart chakra, known as anahata, directs the movement, thereby compromising the electrical and muscular functions of the heart.

Physical manifestations outside one's physical body are mostly derived from the solar plexus chakra, which, in addition to its telekinetic functions, may also promote materialization. Materialization is a process in which extra-physical energy is condensed into physical elements and objects. During this process, the astral and semi-physical auras transfer and absorb a significant amount of etheric fermions, bosons, and hadrons, which, according to Gagnon (2011), are the building blocks of matter. The distinction between physical and non-physical matter lies in the speed of their vibration, which makes astral matter lighter and more expansive in comparison to physical matter. The solar chakras and the heart chakra facilitate the condensation of non-physical particles into physical energy by lowering their speed, therefore resulting in their transformation into physical energy. This process then leads to the agglutination of physical elements. Most of the etheric currents that are condensed into objects are in the form of ectoplasm. Mediums and psychics who engage in spiritual healing by donating or transferring energy may experience a significant loss of these elements. As a result, they may overeat for several days, as their appetite seems to remain unsatisfied even after consuming normal amounts of food. Alternatively, they may remain persistently hungry without overeating. Consequently, it is not uncommon to observe spiritual healers who have either a rather bulky or an extremely slim physique. Usually, the elements that can be created in materialization are small objects, such as leaves, small crystals, or pieces of paper, or even molecules of medicine that can be transferred to water and administered to a patient. In some cases, materialization may also involve fragrances or body cells, particularly in cases of healing or spiritual surgery.

In terms of spiritual surgeries and healing, the sacral chakra, also known as the sexual chakra, plays a significant role in an individual's ability to cure others. When activated, the sacral chakra can promote physical revitalization in living beings. Although this paranormal capacity is rare among the general population of psychics, it is more prevalent among paranormal monks and yogis. The sacral vortex is connected to the body's testes, ovaries, and urinary system, as well as to the electromagnetism produced by the body's fluids. An activated sacral chakra enables an individual to replenish the vital force of living beings through the use of vital energy. However, it is important to note that this energy may benefit many forms of living beings while also potentially harming more ethereal forms of life, such as those of some elementals of nature, due to the contradictory variance of frequency between the energy from the medium and theirs. This is especially true if the medium lives a life of malice and sexual exaggeration.

Dreams and Astral Travel

Of all extrasensory perceptions, dreams are a common occurrence among most individuals. Often perceived as a completely physiological phenomenon and not regarded as exceptional events, they may also offer insights into the realm of the supernatural, particularly in the context of lucid dreams and astral projection.

The examination of dreams from a neurological perspective has been the subject of research by Seligman et al. (1987), who posited that they are the product of the cognitive integration of images and information acquired during an individual's lifetime. Equivalently, the orbitofrontal cortex, located in the frontal lobe, is regarded as the embodiment of the ego or self and serves as the center for an individual's introspective consciousness. This region not only plays a role in the cognitive decision-making process but also functions as the brain's "fact checker" (Rolls et al., 2008). With regards to the limbic system, emotions such as desires and fears can be traced to, among other regions, the amygdala, a group of cells located in the region. This critical center significantly influences the processing of emotional memories, decision-making, and the generation of various emotional responses. During sleep, the orbitofrontal cortex is in a state of rest while the amygdala remains active. As a result, dreams may lack the qualities of the prefrontal cortex, such as its verification of facts, logical decision-making, or control over one's identity. However, the amygdala, where information regarding desires and fears resides, will set the tone for the dreams. Memories, images, sounds, and emotions stored in an individual's brain during their lifetime, particularly recent memories, will surface in dreams as a means for the brain to discharge fears, desires, and over-imagined scenes. The repetition of thoughts creates a neuronal pattern that is likely to be repeated during sleep, thereby resulting in ordinary dreams that serve as the digestion of repetitive thoughts and linked emotions. For instance, a persistent fear of losing one's keys or wallet may result in a dream where the loss of these objects is indeed experienced. In this manner, the brain eliminates the fear by transforming the thoughts into reality. Upon waking, the thought of losing these objects will occur less frequently and with less intensity.

    The experience of dreaming can also have an astral component, where the dream reveals the experiences of the astral body after it separates from the physical body during sleep. When a person falls asleep, their cardiorespiratory system slows down and their brain wave patterns decrease, leading to the relaxation of the etheric body, which serves as a connection between the physical and astral bodies. This reduced tension enables the astral body to detach from the physical body, though energetic filaments still link the two bodies, ensuring that the astral body will not become lost, trapped, or permanently disconnected—which would result in the death of the physical body. When the spirit is free to explore, it can travel beyond the physical body's location in the material plane, although it may occasionally remain close to it if the individual is preoccupied with mundane concerns. During the period of astral detachment, the astral body will often engage in experiences according to its individual preferences. These experiences may include reconnecting with familial relationships from past lives, revisiting favorite locations, participating in spiritual lessons, aiding benevolent spirits in their efforts to help others, and, in some cases, participating in toxic mundane gatherings or even fixating on an enemy. However, these experiences are rarely remembered upon waking, due to the limited awareness during the event, where materialistic matters tend to dominate conscious perception. Despite the reduced awareness during sleep, which affects the clarity and vividness of the experience, the astral body may still retain information from its adventures when it rejoins the physical body, thereby allowing distorted scenes to emerge in a dream. As a result, even when lucidity is limited, the physical brain can retain some of the information and memories.

The physical images, sounds, emotions, and general experiences that are stored in spirits' minds, or in their mental bodies, serve as the columns that support the structure of a dream, just as the brain's serve as the bricks. For example, if the spirit visits a bright, vast location and encounters their mother from a past life, their physical brain will attempt to shape and fill in the vague memory with the closest images, sounds, colors, and emotions that it has stored. By using familiar analogues to reality, the individual may remember a dream of being at a beach with their current life's mother; as a sunny beach is the closest approximation their brain could make to a bright, vast space, and their current mother provides form to their former one. Another example is an astral experience where an individual sees a friend or family member exhibiting behavior that is inconsistent with their typical demeanor in real life. In this instance, the physical brain has never encountered this person before and therefore attempts to identify them by finding similarities in appearance and personality traits with individuals it is familiar with. If the individual dreams they have a close friend whom they have never met, it is likely that the brain failed to associate the memory with anyone they know. Overall, the images and emotions stored in the physical mind—as neural information—tend to give shape to the majority of an individual's astral recollections.

The experience or acknowledgment of astral occurrences during sleep may manifest as symbolic scenes within dreams. Symbolic dreams are often mistakenly regarded as poetic expressions, but in reality, symbolism in astral or mental experiences is not intended to be poetic. Symbolism is commonly utilized as a means of communication across different dimensions, particularly in dreams, where they typically present as scenes and ideas that, when interpreted based on the individual's understanding of those symbols, convey intricate information that can be easily comprehended without inducing shock, confusion, or trauma. In some instances, the symbolic scenes within dreams may be constructed by spirit guides to transmit essential information to the individual's consciousness, but more as implicit ideas than as impactful information. Alternatively, the symbolism may simply be a digestion of images that the physical brain processes as a collection of perplexing scenes.

To differentiate between a purely physiological purgative dream and a dream that is a recollection of astral experiences, two factors should be considered: the length of the dream and the degree of unreality or fantastical elements. Dreams that comprise a series of disconnected scenes, or those that lack coherent sense, or those that feature many images from the individual's recent experiences in real life, are typically purgative dreams. On the other hand, dreams with a longer context or storyline, or those with a sense of individuality and morality, as well as those in which the individual appears on a journey after having been somewhere, are more likely to be astral memories. A further intricacy of astral travel occurs when the physical body, through the conduction of energy threads, receives instinctual impulses from the spirit. For example, during an astral projection in the deep waters of the sea, the traveler's breathing would not be affected. However, if the individual worries about the depth of the water, their conditioned response may cause them to lose their breath during the experience, potentially causing them to awaken in their physical body feeling as though they were breathless.

The psychic experience of astral traveling is a phenomenon where the human spirit is able to detach from the physical body and lucidly explore the astral planes and integrally remember their activities. Different from a dream but sharing with it the fact that it is experienced as the body falls asleep, the process involves the astral body separating from the physical body while one is asleep and remaining connected by energy cords made of ether, a semi-physical substance mostly produced by the chakras. All humans undergo spiritual detachment during sleep, but astral projection is different in that the individual is conscious and aware, enabling them to "travel" and engage in desired activities. During astral projection, the spirit is not entirely free to go or do whatever it pleases. The individual's moral principles and desires will invariably attract and maintain oneself in a particular "layer" of the astral plane. The astral plane is comprised of several levels of reality. Comparing the astral plane to radio waves, it can be understood as several places overlaid that vary from frequency to frequency. The more worldly-driven an individual is, the more attracted they will be to Earth's astral layer, and they will thus perceive astral reality and its visuals as very similar to physical reality. The more altruistic the traveler's intentions, the more likely it is for them to converse with enlightened beings, visit wonderful locales, travel via teleportation, and so on.

In the astral planes, individuals have the opportunity to help suffering spirits receive the care and rescue they need. This is particularly relevant for spirits who are experiencing mental illness, addiction to mundane aspects of life, or are still grappling with bodily decomposition after death due to a sense of unacceptance about their new reality. While spirit guides and benevolent entities possess the capability to aid recently departed spirits and those who remain trapped in a state of psychological death, the significant disparity in frequency between the two types of spirits poses a challenge. Kind beings tend to be too subtle, while suffering spirits tend to stay close to earthly energies. Therefore, the dense energy of an incarnate person appears to be useful and compatible with aiding the suffering spirit. Other individuals may also participate in these rescues while asleep, as their spirit guides prompt them to project energy to those in need and communicate with the recently departed spirits, reassuring them about what they are about to encounter. Recollection of such events is rare and only becomes apparent to astral travelers who possess broad awareness and lucidity. While these experiences may manifest in dreams, they often go unnoticed upon awakening.

The ability to astral project is inherent in all human beings. However, only a limited number of individuals may experience this phenomenon due to their materialistic conditioning and low extraphysical awareness in their everyday lives. To enhance this ability, one must cultivate increased awareness through meditative practices.

Astral travel experiences can be remembered similarly to how vivid dreams are recalled. However, the physical brain may struggle to process and interpret the information gathered during the astral experience, leading to incomplete or forgotten memories. This is because the physical brain has limited capacity to store vast amounts of information and may not be able to assimilate the subtle data from the mind as it would with information obtained through the physical senses. In some instances, full or partial memories of astral travel may not be retained after waking.

Occasionally, sleep paralysis may precede an episode of astral projection. As noted by Sharpless (2016), sleep paralysis is characterized by conscious awareness and complete muscle immobilization, occurring during the transition between wakefulness and sleep or vice versa. According to Avidan et al. (2011), sleep paralysis affects between 8 and 50% of individuals at some point in their lives, with 5% having regular episodes, affecting males and females equally. During sleep paralysis, individuals may experience hallucinations that can cause fear. Despite Avidan et al. attributing sleep paralysis to genetic predispositions or sleep disorders, the root cause remains unclear in the fields of neuropsychology and sleep biology, as acknowledged by the UK's National Health System (2018). The mechanics of sleep paralysis stem from the relaxation of the brain and cardiorespiratory system, which allows the etheric body to relax and its magnetic properties to decrease. If the semi-physical aura extends too far, the detached astral body may still be influenced by the physical body, even though consciousness is disengaged from the physical brain. The state of paralysis occurs when the astral body tries to move a physical body that it is not enveloped by but rather just overlaid on. Fear activates the respiratory and circulatory systems, thus releasing adrenaline into the bloodstream and thus tightening the etheric body, which makes the astral body reattach to the physical body. In a similar vein, hypnic jerks, also known as sleep twitches, as described by Sathe et al. (2015), are benign myoclonic flicks that usually occur on falling asleep and during rapid eye movement (REM). They are observed by sudden jumps moments after one falls asleep and often feel as though one is falling. While scientific studies suggest that hypnic jerks may be caused by caffeine, nicotine, or sleep deprivation, the exact cause of this phenomenon is still unclear. However, from a spiritual perspective, the primary cause of sleep jerks is related to the separation of the astral body from the physical body. The interference and magnetism of the physical body may still attract the astral body back to the body by means of the etheric body, an incredibly strong magnet that is most potent during wakefulness. When the spiritual body is abruptly returned to the physical body, the person may feel as if they have fallen or passed through their lying body. It is crucial to note that while scientific studies may show possible causes for physical phenomena, the primary reasons for countless phenomena are of a spiritual nature, with the resulting experience being in the physical body in the material universe.

To experience astral projection during sleep paralysis, it is recommended to remain calm and avoid movement, eye opening, or speaking. Relaxation and visualization of the auric body expanding prior to bedtime may facilitate the occurrence of sleep paralysis and, thus, facilitate an astral projection. During sleep paralysis or just before leaving the physical body, it is common that the individual experience noises inside the skull. These noises may sound like metallic bars hitting the ground or the rumble of a train. However, these sounds are not physical in nature. Rather, they represent the human capacity to interpret the change in frequency that occurs when a person transitions from the physical plane to the astral plane. Occasionally, words or a person's name may also be heard, such as shouts or curses.

However, these are not usually the result of malicious spirits. Instead, they are often mental emanations from the individual perceiving objects in their own aura.

Conversely, lucid dreaming is defined as a continuum of awareness (Barrett 1992). By blending aspects of dreaming and astral travel, lucid dreamers can actively participate in their dreams, manipulate their environment, and influence the direction of their dreams. However, as the dream unfolds, the dreamer's level of alertness typically declines significantly, resulting in dream scenes that are often disjointed and illogical. In essence, lucid dreaming is a form of astral travel that the physical brain primarily interprets as visual imagery.

Meditation and the Unlocking of Potentials

While mediumship and psychic abilities are thought to be innate and can only be fully developed in people who are born with them, there are ways to partially uncover and develop extraphysical sensitivity through exercises like meditation. Meditative states are characterized by specific patterns of brain activity, known as brain oscillations or brainwaves. A study conducted by Aftanas and Golocheikine (2002) found that these patterns typically range from 13 to 8 cycles per second, which corresponds to alpha brainwaves. However, deeper meditative states are associated with even slower brainwaves, specifically theta waves, as noted by Cahn and Polich (2006). Thus, the brainwaves observed during meditation can provide valuable insight into the depth and intensity of a meditative experience. As explained by Einstein (1905), matter is energy in a condensed form, while energy is matter in its radiating aspect. In this fashion, it can be assumed that the human body is composed of energy, which naturally generates physical radiation.

The physical aura, produced by the human body, is a radiating field that interacts with other fields in a natural physical style. Similarly, the seven main chakras are where the exchange of semi-physical energy is most intense. Wagner Borges (2005) discusses how intentionally visualizing a pulsing light onto the third eye chakra promotes lucidity as well as the expansion of that semi-physical vortex, implying that its radiant emanation acts in a more expansive radius during meditation. This expansion can enhance clairvoyance, the ability to see beyond the physical realm, and even perceive other entities. The reason behind the visualization of light is that one's mental creations, which do not pertain to the physical brain, have the capacity of creating and molding light onto one's own reality. What appears to be pure imagination of light and color is actually the creation of these entities in the etheric, astral, and mental planes.

The hypnagogic state is a transitional state between wakefulness and sleep that can be experienced during meditation. This state is characterized by relaxed brainwave patterns, which enable communication between the conscious and unconscious minds. By bypassing the brain's rational filter, this communication can facilitate the processing of extrasensory information, leading to auric expansion, psychic experiences, intuitive insights, and memory retrieval. These effects can ultimately enhance perceptions of physical and nonphysical objects and places. With regular meditation practice, the vortices associated with extraphysical perceptions become increasingly active, resulting in expanded states that may persist for days after meditation. This suggests that psychic and mediumship episodes may occur spontaneously, without any additional preparatory meditation.

Meditation can be understood as a practice that involves

assuming specific bodily postures and achieving a state of mental stillness. However, the true essence of meditation lies in directing one's focus towards what is truly important, specifically towards one's inner self. When an individual chooses to engage in meditation, they consciously decide to divert their attention away from external and physical distractions and instead concentrate on what occurs within. The perception that is absorbed by one's aura is present in the mind, and by concentrating on it, one can decode and comprehend the extraphysical information that has always been present, despite having never been comprehended. To fully unlock their psychic or mediumistic abilities, an individual must remain open to receiving and facilitating the unfolding of additional information, particularly when visualizing light. The more light that one can gather within their etheric and astral body, the more expansive and profound their spiritual experiences can become. By envisioning light radiating upon each of their primary chakras, one can expect their state of being to remain active throughout the day, thereby increasing their extraphysical functions and spiritual capabilities beyond their physical influence.

While meditation can assist in relaxation and enhance cerebral structure, its primary benefits are related to one's spiritual body and energetic fields. By simply picturing light falling from above, as though in a shower, an individual can cleanse their auras and energize their astral bodies, which will, in turn, enhance their perceptions. This same visualization can also be applied to cleanse and prepare one's environment for spiritual rituals, prayer, healing sessions, or general well-being.

After a period of meditation, elevated spirits may imbue the individual with intense feelings of happiness or other beneficial emotions. As the individual carries these feelings into their daily life and interacts with others, they may radiate an aura of joy that can be passed on to those in need. This can be seen as a form of spiritual transmission, in which benevolent forces utilize the person's body to perform acts of kindness.

Extraphysical sensitivity requires a high degree of morality and compassion, as there are countless entities that may attempt to deceive an individual by assuming false appearances and using deceptive language. Such entities may pretend to be ascended masters, deceased individuals, or other entities they are not. This underscores the importance of humility and caution in mediumship and psychic practice. While it is possible for ordinary individuals to increase their sensitivity and communicate with the extraphysical world, they must also accept the responsibilities that come with such abilities. Indeed, those who exhibit arrogance or vanity in their abilities may be subject to karmic consequences, not for being deceived by the spirits, but for their hubris in believing they possess superhuman powers.

Overall, meditation can enhance mediumship, but for proper development of this ability, it is advisable to seek guidance from a temple that specializes in mediumship. Such temples are protected places where a group of spirits typically prevent the penetration of negative entities. It is important to note that psychic development and mediumship development are distinct aspects of spiritual growth that can be cultivated through meditation and other exercises. While psychic development involves enhancing extrasensory perception and abilities, mediumship is a natural capacity for mediating between different planes of reality. Only those born with a natural talent for mediumship will be able to fully experience this ability.

PART 3

Low Magic and Ritual Practices

Witchcraft involves the manipulation of energy to achieve desired outcomes in both the physical and astral planes. While some may employ these practices for personal gain, the law of karma operates in tandem with it, and hidden or occult entities may participate in the rituals, regardless of the witch's awareness. This chapter explores these concepts and also delves into the notion of spiritual contracts and treaties that are common in low magic, where practitioners bargain with entities to obtain favors (it is worth noting that this chapter aims to elucidate the rituals and elements of witchcraft but is not to provide instructions or guidance to readers on performing such practices, as they are not recommended).

Additionally, while potentially sensitive subjects are addressed, the chapter describes why prayer is the most effective means of achieving desired outcomes.

Witchcraft Spells, Voodoo, and Incantation

The energies that are manipulated in a ritual, including the elements used to produce physical phenomena, can be obtained from different sources, such as offerings, the fluids of the magician's own etheric body, or the cosmic ether, also known as prana. The energies obtained from an offering are the etheric doubles of the food given. These energies are the radiant bodies, or auras, of the foods, generated by the movement of their physical particles. The magician's own etheric energies, which flow through meridians and serve to vitalize the physical body with semi-physical light, can be used in magic as the ingredient that nourishes a thought-form. On the other hand, cosmic prana is essentially a cosmic type of light that invigorates both organic and inorganic molecules. Prana can be of various natures, such as solar prana from the sun, angelic prana from groups of benevolent spirits, and nature prana emanating ceaselessly from natural spots.

The first step in creating a spell is to construct a thought-form, which is essentially a holographic image that the mind produces and externalizes. These thought-forms can take the form of objects, sensations, or abstract ideas, such as scenes from a possible future, and are commonly included in witchcraft rituals and other energy manipulation practices because they serve as the vessel for the energies obtained to manifest the desired outcome. The externalization of semi-physical photons and plasma produced in the spiritual brain is required for the creation of thought-forms. These photos are naturally injected into a miniature copy of the original thought produced. The objects or ideas typically disappear if the cerebral thoughts cease to be reproduced. These objects can only be magnetized by physical bodies, that is, by the individual who is incarnate. Although thought-forms can be produced by discarnate spirits, those objects would mostly remain in the astral realms and produce astral effects. Conversely, if the thought objects are magnetized with semi-physical ether, that is, semi-physical energy, they are more likely to condense or unfold in the etheric and physical planes. For instance, a simple thought of a pair of shoes generates a thought-form of those shoes, which usually appears hovering above the person who generated the thought. Recently generated thought-forms are akin to pencil sketches on paper, lacking in color and having a translucent appearance. As the thought recurs progressively, the thought-form begins to take on an appropriate form, and the shapes become more vivid, detailed, and seemingly three-dimensional. Over time, they appear to take on a life of their own, emanating their own aura and moving naturally. These holographic objects created by the mind are bound to the neocortex by magnetic filaments, particularly in areas such as the frontal lobe. They are kept active by millions of strands directly connected to neuronal synapses.

When a thought-form is highly magnetized, it can influence other similar thought-forms in the universe, creating a chain of resonant thoughts. This can lead to the manifestation of those thoughts into physical reality. To increase the effectiveness of a ritual aimed at manifesting a specific object, such as a car, it is important to use physical elements that possess qualities similar to those of the intended object's etheric doubles. For instance, a ritual aimed at acquiring a car might include elements such as a miniature or symbol of a car, metal keys, and root vegetables. Similarly, if the desired outcome is to manifest passion, using red fruits in the ritual may be more appropriate.

Direct control of energies is another method that can be used to condense subtle energies into physical reality. The manipulation of energies in offerings can be understood through analogies derived from quantum physics phenomena. Some theories that can be used for didactic purposes are:

- Wave-particle duality: According to Einstein et al. (1938), every particle or quantum entity can be thought of as either a wave or a particle;
- Wave function collapse: As analyzed by Griffiths et al. (2005), the wave function collapse occurs when a mathematical description of the quantum state of an isolated quantum system reduces to a single state as a result of interaction with the outside world;
- Quantum superposition: This is a fundamental principle of quantum mechanics that states that any two (or more) quantum states can be added together and "superposed", and the result will be another valid quantum state. Every quantum state, on the other hand, can be represented as a sum of two or more other distinct states. (For a deeper understanding of quantum superposition, see

Dirac, 1947);

- Quantum entanglement: A phenomenon that arises when a set of particles are produced, interact with each other, or coexist in close proximity in such a way that the quantum state of each individual particle cannot be independently described without reference to the state of the others, even when these particles are separated by a significant distance. The first theories on quantum entanglement appeared in 1935, when Einstein, Podolsky, and Rosen wrote for the Institute for Advanced Study in Princeton, New Jersey.

- When a thought-form is imbued with the energy of carefully chosen and manipulated elements from offerings or one's own energy reserves, it begins to connect with its counterpart in the material dimension. For example, a thought-form of a car starts to seek its condensed counterpart, becoming a future etheric double of that car, akin to what occurs in quantum superposition. Similar molecules attract each other, so the magnetism of the car's etheric double as a thought-form attracts the physical molecules of an actual car. This molds probable future occurrences to manifest a physical car for the individual who initiated the thought-form. The magnetized thought-form does not assemble a car but rather connects the physical car to itself, as physical cars have a frequency closest to the magnetized form.

Spirits, especially those incarnated, appear to be what can be understood as consciousness magnets. But the magnetism may only act in favor of one's wishes if they actively exercise thoughts accordingly. When the etheric doubles of offerings are used, the non-physical thought-form becomes semi-physical and magnetic. However, the magnetism from one's aura is still an important part of the magnetized thought-form. In other words, even if a thought-form is created and magnetized using the etheric double of an offering, an individual's emotions are still necessary to direct the energy from the offering into the hologram. If the desired outcome was an event or situation rather than a material object, the magnetized and molded probable future occurrences would become magnetic enough to override any other future occurrence, thus making the wish a reality. But nevertheless, wishes, petitions, and rituals will not override karmic laws. That is, an individual who is to karmically experience poverty for such a period due to a past of arrogant greed and avarice will hardly achieve large sums of money through spells, no matter how complex they seem to be. In like manner, a woman who has aborted several times in a previous life may not be able to generate a child with the aid of witchcraft either, because energy in her spirit's womb cannot flow freely, so holding another soul would encounter severe difficulties. It is crucial to remember that these karmas are commonly chosen by the individuals themselves, upon guilt or resolutions for lessons in the physical plane.

In witchcraft works involving another individual, as in a binding love spell, for example, the magnetism present in the target's astral body is connected to the other individual's body through condensed light currents. These wires are connected to their respective vortex centers located in the navel region of both individuals and to the chakras near the adrenals and gonads, which are glands that concentrate considerable electromagnetism. These ligaments are comprised of a variety of condensed photons and plasma and can be found in all kinds of relationships between people who know one another personally. For these wires to be sustained, the energies of those vortices must be drained, which invariably causes the two individuals to lose vitality, therefore making them prone to diseases, accidents, negative moods, and spiritual vampirism. In this kind of lower magic ritual, the two individuals create greater affectionate bonds, albeit ones that are more addictive and vicious in nature than those from a loving perspective. Love or passion cannot be replicated in love spells; instead, they cause the victim to feel deluded and uneasy in the absence of the other. In love-binding rituals, subatomic currents generate modulations in the microtubules of the victim's neurotransmitters, altering their brain biochemistry and building new neurotransmitters that shape the inexplicable addiction to being around someone, even though the presence of love is nonexistent. Serotonin, endorphin, dopamine, cathecolamines, and especially oxytocin levels are modified in this way, revealing the spell to cause rather physical results.

In the ancient spells of sewing someone's name into a frog's mouth, this would be mirrored as an example of superposition and entanglement, where the frog becomes the paper and the paper becomes the frog. In this sense, a particle (like paper or a frog) is also a wave and thus contains information. The information in the paper has a vibratory signature that matches the name of the victim and its matrix. The frequency of someone is commonly known as their "energetic signature." Quantum entanglement is the best analogy regarding this variety of magic, which can also be applied to voodoo doll rituals. When the ritual enchanter prints the victim's mental image on that piece of paper, the paper's field automatically conduits the victim's aura. With the suffering of the frog, its electromagnetic field decreases in frequency and increases in condensed light, that is, light traveling at a much slower speed. The etheric emanations of the paper, hence, interact with the frog's aura waves, generating a third wave similar to what occurs in a quantum superposition. Since the paper is connected to a victim, the other end of the wire—the victim—begins to experience the same frequency changes as the first end, that is, the frog. In a nutshell, a frog "here" affects the victim "there," a photograph "here" affects the victim "there," and a candle "here" affects the victim "there." A minor differentiation between these spells and the voodoo doll ritual is that in the above spells, there may or may not be similarities with quantum entanglement. Energy can simply be directed to an electromagnetic field analogous to that of the object used. But in a hoodoo ritual, there is quantum entanglement and subsequent replication of the electromagnetic field of a person that is projected to be the doll's. The idea is that it works as though the doll's electromagnetic fields are the victim's etheric double.

The explanation of lower magic spells involves understanding the scientific principles behind the events they seek to generate. However, this book does not aim to delve into the many intricate details required to create a ritual that mimics particle-wave entanglement or superposition. The mere act of putting a piece of paper in a frog's mouth without an extra-physical activation is likely to yield no extraphysical results. Furthermore, with or without efficacy, the karmic repercussions of such deeds, when magically activated, may compromise years of the aggressor's existence.

One form of harmful ritual involves a sorcerer employing mental techniques to gain access to the astral body of a sleeping victim. By enlisting the aid of a spirit or group of spirits, the sorcerer can retrieve traumatic memories from the victim's past lives. By enlarging the connection between the victim's current personality and their old personality, the spirits bring dormant memories into the present, causing disturbances in the victim's current life. It is important to note that this type of access to mental fragments is only possible if the victim has a karmic debt from using magic against others in the past. The process of undergoing such magic can therefore serve as a means of spiritual cleansing and debt repayment, leading to sublimation.

Another practice that uses long-distance ritual and is

common in Vodun is that of voodoo dolls. Vodun, a religion that places great importance on the power of amulets, believes them to be stronger than spirits and even capable of controlling and dominating them. The doll represents the person, and the pins are the forces that are implanted into the person's etheric body. Small magnetized thought-forms in the form of suggestions are designed to produce a desired outcome, such as tissue regeneration or degeneration, and are intended to penetrate the doll in unison with the needle. Simply put, the thought-forms imbue the needles with the desired intention and serve as a vehicle to penetrate the doll, and thus the patient's or victim's etheric double. Through its connection to the etheric double of the person whose doll it represents, the doll is able to obtain the magnetized thought-forms for itself. In the case of a malignant doll, a ritual is developed in which trance is occasionally followed by the slaughter of animals for offering purposes, allowing nefarious spirits to absorb their share of physical sensations while also manipulating magnetism to conduit the doll to the victim. While the practice of using a doll or effigy made of any material and piercing it with needles may seem like a means to harm someone, it is actually a practice more commonly used for healing purposes, both for oneself and others. The roots of voodoo extend far beyond such malevolent practices, as seen in the nkisi tradition of Benin (Cuoco, 2014). Here, the voodoo doll serves as an amulet for health, prosperity, and protection. During a benevolent ritual, the practitioner chants in order to achieve trance states, whereby they fixate their gaze on the dolls while visualizing the individual superimposed on the doll. Thus, while vodun dolls can be used to harm others, they can also be used as instruments for connecting to the body and producing positive effects.

To understand how voodoo dolls can affect their human counterparts, it is important to look at how one object can affect another from afar. The connection between the amulet and the person is made through the person's clothes, hair, or other personal items that are woven into or put into the doll. Thought waves from the victim (or patient) are naturally stored in the body's cells and on personal items. A trained psychic is normally able to tell who had the most contact with an object using psychometric methods, which are perceptions of the mental impressions that an object or place has had. Auric waves project layers of ether onto anything they come into contact with. Those ethers have a unique signature. A signature, in these cases, indicates who externalized those ethers. Like DNA, the waves emanated by an individual will always carry all the information about the individual who produced that wave. Because the object is inextricably linked to that incarnated spirit, it can be re-magnetized to the individual if a ritual has the ability to produce such magnetism between the two. Gazing upon the doll, in which the sorcerer visualizes the individual to be represented by it, allows the victim's thought-form to resonate with their own hair or other object used in the doll's confection. However, the simple visualization or confection of such a doll will not produce any effects if it is not ritually activated, which is not revealed or recommended in this book.

If the ritual is aimed at someone else's misfortunes, maleficent entities would most likely take part in the ritual, albeit not always revealing themselves to the enchanter. Any harm directed towards others leaves an imprint on one's astral body, akin to an x-ray. This imprint manifests as a miniature sign of the harmful act as it passes through the mental and astral bodies as soon as such a thought is accompanied by a regrettable emotion. Likewise, positive actions also leave a mark on one's astral bodies. Generally, individuals who employ voodoo dolls or any witchcraft practice with the intent of causing harm to others often face karmic repercussions, such as prolonged illness in the organs that were targeted by the pins, in the case of voodoo dolls, or inability to find a spouse, in the case of love-biding spells. It is essential to note, however, that someone who falls victim to magic may have used witchcraft against others in a previous existence.

Pacts and Spiritual Contracts

Spiritual contracts are purported arrangements or agreements that individuals may make with spirits for specific services. These agreements supposedly cover a wide range of desires, including wealth, fame, and personal magnetism. While some people may believe that they can sell their souls, it is important to note that such arrangements are not everlasting, as spiritual evolution must inevitably occur for both the client and the spirits at some point in their existence. What is purportedly sold is one's servitude in the afterlife. Pacts with these entities can range from 1 to 7 years, and in rare cases, the entire duration of one's life if the client is able to persuade others to join the contract as well. Once the contract is completed, the person is not necessarily expected to lose their wealth or fame, but they will no longer receive further assistance and may need to gather their own resources if they wish to continue reaping the benefits of what was provided to them.

A practitioner of the magical arts acts as an intermediary between a malevolent spirit and the client and is responsible for conducting the ritual and communicating its terms to the client. These individuals charge for their services as mere mediators between the parties, but they also bear the karmic consequences of their dealings with the entities upon their own passing. Once the contract terms are agreed upon, the spirit enlists its servant spirits to help the client achieve their desired outcome. The spirit responsible for an individual's soul contract is usually a representative of a higher-ranking spirit, which means that the person may end up becoming a slave to other spirits on the spiritual planes. Often, these spirits are deceptive and may suggest that the person's service will be mild on the other side. These servant spirits are often enslaved themselves and typically work by influencing others via suggestion and subtle energetic methods, such as implanting devices into people's astral bodies, to lead them to act or think in accordance with the desired outcome. They may also hinder the energy flow of potential rivals of the client, making it easier for the client to achieve their goals.

Even if the pact aims to do no harm to others, it is inherently apparent that it harms oneself, as an agreement to serve as a slave goes against universal laws. Moreover, the actual service that the person is expected to provide is often much more significant than what was initially agreed upon. In many cases, this service may include fulfilling demands made by other incarnate humans in their own ritual agreements, such as causing mental illness in others or inciting violence and addiction. Furthermore, as a slave for malevolent forces, one is likely to be demanded to cause harm and destruction, which is essentially agreed upon when signing a spiritual contract. In this regard, aspiring to thrive in life is distinct from dedicating oneself solely to mundane status and power, especially if they agree to be part of a spiritual group that focuses on making others' lives worse. As the client accumulates a spiritual debt through their connections with these entities, the spirit can temporarily enslave the former client upon their physical passing, in accordance with the law of cause and effect. Therefore, selling one's soul implies agreeing to be permanently under the rule of that spirit after death.

Although such an arrangement may only last for a limited period, it could potentially extend for centuries, as time on other planes does not parallel the physical plane of

Earth. Additionally, the individual may accumulate further karmic debt for any actions they undertake while working for these spirits. The karmic debts that individuals incur upon selling their souls are substantial and enduring. They will not only be subject to the control of malevolent spirits but may also encounter these entities in future reincarnations. This is one of the reasons why many people have been under the influence of malevolent spirits for centuries. Although not all individuals with long-standing connections to spirits have sold their souls, they may have committed to assisting these spirits while in the lower astral realms. These ties persist even after reincarnation and can lead to ongoing spiritual consequences until the laws of cause and effect are restored. When their karmic load and lessons permit, benevolent spirits frequently approach these people who have been in slavery for a considerable period of time, provided they also exhibit a profound sense of regret. Regret, in this context, implies a sense of attaching value to what was once disregarded, namely their previous incarnated life where they valued wealth, power, or passions over the opportunities to forgive, help, and learn in the physical realm.

Offerings

Offerings made to spirits, deities, gods, or nature are intended to assist with a request or enhance the relationship between the person making the offering and the entity receiving it. During a ritual, when specific foods or objects are presented to a deity or spirit, the offerer's aura and the offering are drawn together. This serves to prevent other nearby spirits from claiming the offering for themselves. When an offering is made, various elements such as chants, candles, incense, symbols, or prayers may be employed to facilitate the loosening of the elements in it, enabling the deity to readily access its etheric double. The rhythm of the chants, the color of the candle, and the herbs used in the incense may all be energetically similar to the essence of that particular spirit or deity, at least in terms of the sort of emanation they reveal.

It is a common misconception that the spirits to whom offerings are made actually eat and drink what is offered. The entity, or spirit, offered purely manipulates the etheric double of the elements "given" to them, as long as those foods and objects are resonant with the frequency they operate in. For example, in a ritual where the offerer wishes to please or spiritually align with a certain deity known as the "deity of beauty and love," the offering to be given would need to contain components that emit vibratory emanations similar to the deity's nature; that is, the elements present in the offering must have been conceived from the same source frequencies that that deity is analogous to or works under. Even entities with no ties to any particular god or supreme being will only receive offerings that are compatible with their frequency. Not only is personal preference a factor, but also the difficulty that one may encounter when attempting to manipulate the etheric double of elements that can be described as either "faster" or "slower" than what they are accustomed to. In this case of the deity of love and beauty, the elements offered may include, but not be limited to, apples, sweet beverages, juicy fruits, fragrant incense, red, pink, or yellow flowers, as well as make-up, perfumes, mirrors, and delicate decoration, depending on the tradition, culture, or religion in question. In this way, a spirit representing the deity can use the etheric double of liquids, sugars, natural pigments, and objects as thought-forms to produce something resembling love or beauty. Along these lines, giving such a deity an offering containing whiskey, peanuts, and violet candles may contradict their vibratory nature and hence the vibration of love or beauty, which is analogous to water, sweetness, and the color pink—which whiskey, peanuts, and violet candles are not.

In rituals where the offering is used to worship a divine being or to gift a specific deity, a spirit assigned to that deity's group may manipulate and refine the etheric double of the offering, subsequently transferring the energies to the individual's auric field and thus the spiritual influence the offerer expects to receive back from the deity. For this reason, sourcing the right elements for any ritual offering is immensely necessary. In order to determine what the frequency of such an element is and what its vibration is, it is crucial to understand its physiologic, botanical, and chemical components. Generally, their color indicates the frequency of their etheric double. Reds are associated with agility, vitality, conquering, and sexual energy; thus, red fruits and drinks, as well as red objects, almost certainly connote one or more of these qualities. In this manner, the color red is mostly associated with spiritual entities that play a role in self-esteem, desire, and karma, whereby they utilize organic pigments and light reflected as color to produce extraphysical objects. Nevertheless, the quantity of water present in the elements also reveals the temperament of the spirit, that is, their field of work. Red fruits high in water, like watermelons and apples, will rather assuredly be related to entities dedicated to love matters, family, and relationships. If their pulp is red but not their peel, the fruit is more suitable for lack of desire petitions; if the peel is red but not their pulp, this fruit is more suitable for wishes regarding situations that lack or need emotional interest. Red fruits with a low water content, such as chili peppers or dried goji berries, are typically associated with vitality and enthusiasm. The same principle holds true for beverages: the more alcoholic they are, the longer their oscillation will be. In the same vein, alcohol may be used in offerings to deities who work in the cleansing of places and auras or in protecting or guarding specific places and people. The nature of the beverage's component, such as malt, vegetables, barley, fruits, or herbs, may also determine its corresponding use. As such, juices and water relate to entities that work with emotions, relationships, family cases, self-esteem, and harmony between individuals. The sweeter the beverage, the more related it is to love, as fructose and its chemical subdivisions are also collaborators with joy, passion, and euphoria. Orange-colored foods and objects are invariably associated with movement and enthusiasm. This correlation is based on the travel speed of the element's etheric photons. The red light is the fastest traveler, whereas the violet light is the slowest. Orange is the second-fastest color in this case; yellow, on the other hand, relates to self-esteem, wealth, happiness, and beauty; green invokes work, studies, health, and growth; blue relates to mental clarity and a lengthy perspective; and violet is linked to purification and change.

The pigments and water content of the offering should be the best at indicating its etheric attributes; however, the shapes and forms of the elements displayed also represent an important method when sourcing elements for an offering. Pointy leaves and sharp-edged fruits indicate high velocity, as well as the end of cycles. Entities that work with aura cleansing frequently use round fruits and leaves. Furthermore, when the element is soft to the touch, it may indicate the nature of the entity associated with it. Velvety elements are related to the color white, even when the element itself is of a different hue, and they relate to peace, faith, and ascension; hard shells and thick, dry leaves are related to the colors brown, dark green, and red, indicating justice, law, and progress. Overall, the entire aspect of the foods and objects present in the offering is to be taken into account based on the nature of the entity to be offered and the classification of what is requested in the ritual.

Through chanting and prayer, the etheric double of the elements is volatilized, and the chanter imbues the offering with their intention of giving to a deity. As the offerer expands their aura onto the offering, the deity's energetic signature is imprinted on it, allowing the etheric double of the offering to loosen and be more easily absorbed by the entity, typically a spirit representing the deity. When the devotee's aura and the etheric double are in the same auric field, the offering's etheric double is separated from its physical counterpart and absorbed by the entity through a mental command, resembling luminous smoke or radiant steam that leaves the offering. If the petition is accepted, the spirit in charge of the offering will shape that energy and often transform it into a thought-form after absorbing its semi-physical emanation. Thought-forms are a key component of the "probable future," where future occurrences unfold in holographic form before they manifest as physical reality. The thought-form created by the entity is linked to the offerer, making it their own, and the entity contributes with its magnetism so that the thought-form can develop into an actual probable future and become a reality.

The spirit will only proceed with the ritual if their superiors approve it. Benign entities typically base their decisions on the laws of cause and effect to determine whether a desire should be fulfilled or not. Malignant entities may proceed with the ritual anyway, but the outcomes may dissolve quickly as universal laws do not depend on spirits or deities to exist; these laws work ceaselessly for all. If a malignant desire to affect another discovers that the victim was a malignant sorcerer in the past, the works proceed to harm them, aligning perfectly with the laws of cause and effect.

Prayer and Idols

According to various traditions, prayer is a request made to higher realms, where a deity or other entity can hear the request and act on the person's behalf. During prayer, a person's spiritual aura elevates in vibration as they naturally raise their thoughts to something considered holy or transcendental, thus imprinting virtuous outcomes on their own souls. Even if the individual is merely imagining an entity about whose existence they are not sure, they normally still believe that the event they want to materialize has a greater possibility of occurring because such an entity or perhaps "something out there" has more means to make it so.

During prayer, people externalize a thought-form that almost certainly reaches a destination, depending on the intensity of their emotions. Prayers, which customarily take the form of a block of thought, are telepathically captured by spirits, who forward the petitions to spiritual regions where a hierarchical body decides on the best way to assist. Pleas filled with wrath, irritation, or the desire to see others suffer would likely reach the maladjusted spirits instead, who benefit from these malignant energies generated by the magnetized thoughts of hopelessness and anguish. This may result in spiritual vampirism, in which the disembodied spirit encourages the incarnate human to produce more of those thoughts and behaviors, thereby taking more of that life energy imbued with thoughts of vengeance.

Prayer requests are often accompanied by visualization, and this process naturally generates thought-forms that align with the petitioner's desires. Although the thought-forms remain in "data mode," rather than a combination of worded requests and images, they are collected by the spirit as a mental report. This report is then passed on to a more enlightened spirit, who makes decisions about the outcome based on the information contained in the thought-forms as well as the petitioner's merits and karmic consequences. If a prayer request conflicts with a person's life lesson or interferes with the free will of others, it may not be granted, as this would impede their natural learning and development. For example, if a person is in a karmic purgative state, a prayer for healing may not manifest, although the prayer is still stored for future assistance. Similarly, if a person is supposed to experience poverty as part of their life lesson, a prayer for wealth may not be granted, as it would contradict their need to learn from hardship. Ultimately, while the spirits may use prayer to assist a petitioner, they do so within the context of their larger spiritual path and do not interfere with their natural evolution.

After a prayer request is accepted, any potential negative future events may be eliminated, which can positively impact the petitioner's experience. Additionally, other potential future events may be created as thought-forms, linked to the individual, and thereby attract positive outcomes. Spirits, either incarnate or discarnate, may also be influenced to aid in achieving the desired outcome. Even if the request is not fulfilled, it is still evaluated and may assist the individual in ways they may not recognize. All prayers are taken into consideration and are never completely disregarded, including those made by young children and even those with significant karmic debts, such as genocide dictators.

In some cases, individuals may pray for the healing or rescue of another person who ultimately passes away, leading the person who prayed to believe that their prayers went unanswered. However, prayers are always heard, and in those cases, a prayer may have been answered, in which the diseased person was indeed delivered from the pains and found themselves aided in the spiritual planes. As said, all prayers are answered; what differs is the time frame or the degree of manifestation, but in situations where the petition claims for a sports team to win a match or for one's acne to clear in time for a party, these prayers are promptly abandoned once they reach higher planes.

If the prayer is intended to reach God or an exalted entity, such as Jesus, Krishna, or Buddha, the projected wish is indeed assigned to their respective frequencies, indicating that the thought created upon prayer contains a signal for the deity to which the request is directed. The deities would not respond to prayer themselves (although they would be aware of it), but good-natured spirits who work for "the light" and who are usually adepts in tune with those celestial frequencies would. Moreover, if praying to, for example, Saint George, the benevolent spirit to help may naturally be from a group of Tibetan monks or Native American shamans. Spirits are aligned with a certain practice, but they are not religious, as are people on Earth. Furthermore, if one finds themselves as Catholics or Hindus in this lifetime, they should remember that in previous lives, they were perhaps of a different culture and most likely of a different religion.

Overall, prayer is the most effective way to receive support from divine providence through the direct intervention of kind spirits. A single prayer, when made wholeheartedly, is stronger than any magic ritual or incantation; additionally, prayer can even help one alleviate or reduce their karmic accrual in certain circumstances.

Prayer is widely recognized as a powerful means of obtaining spiritual, mental, and physical assistance, and many cultures practice prayer in the presence of images, as if the divine or spiritual entities were embodied in the idols themselves. Although images of deities are not essential for prayers to be heard, they may facilitate the process and help align the individual with the entity's vibrations.

The images and statues of a saint, a god, or any other entity or guardian spirit can act as a channel for the entity's energy and vibrations. As an individual focuses on the object in prayer or during a ritual, their aura begins to couple with the object's aura, which initially consists of a basic magnetic field. The individual's aura then nourishes the object's aura, imbuing it with subtle elements such as etheric force. In addition to donating semi-physical energy, the individual also modulates the energy according to the nature of the symbolic entity they are focusing on. The transfer of etheric constituents commonly occurs in the form of magnetic currents of photons, which are essentially packets of light, but which the human eye cannot perceive. As supported by research, humans are bioluminescent and emit photons (Edwards et al., 1990; Kobayashi et al., 2009). While the amount of visible light emitted is lower than what can be detected by the human eye, there is sufficient evidence that the physical body not only produces electricity (Connors et al., 2004), but also physical light. Along these lines, the efficacy of the extraphysical conduit relies on the upkeep of the etheric currents that sustain it. If devotees cease to transmit photons and magnetism to the effigy, it may experience a reduction in its ability to function as an antenna, albeit not entirely.

Even if the object is activated by a specific individual, others can also use the sacred image as a means of connecting with that particular spiritual frequency. Unlike an offering, which stays connected to the offerer until its etheric double is completely absorbed by a spirit, a sacred object is nourished by one but is not enveloped by barriers between the worshiper and the object itself. An offering, in these terms, is not to function as an antenna, whereas the sacred object's main role is that of a conduit and emitter of semi-physical waves onto the physical plane.

When one prays to the deity represented by a sacred object, a beam of light connects the individual to the image. As an antenna, the image then transmits the message to the spirits associated with that deity or symbolic figure. The entities can receive the message and may also transfer spiritual energy through the object itself. Similarly, lighting a candle in front of a sacred image enhances its ability to capture and convey semi-physical information by emitting visible and invisible particles of light. This interaction strengthens the object's auric field and promotes the transmission of extraphysical fluids.

PART 4

DIVINATION, ORACLES AND TALISMANS

Magic Symbols

A talisman is an object that is believed to possess magical or miraculous powers to protect someone or attract something. Amulets, on the other hand, are a type of talisman that is usually worn close to the body, such as a pendant. Typically, talismans are objects that are ritually created, such as a collar or pouch, while amulets are often of natural origin, such as crystals, pieces of wood, bones, and seeds.

To understand what gives an object the power to attract or repel a specific desire, it is crucial to comprehend the symbols and sigils it may contain. Ritualistic seals, such as symbols and sigils, are rough sketches of the vibratory signature that a particular metaphysical group of entities, egregore, or object radiates. For example, crosses emit a vibratory signature of the union between the spiritual and the mundane, with their radiating filaments crossing and acting on the four poles of the mind. In this case, a cross represents the spirit permeating earthly life. When used as a symbol, swords emit an orderly vibration in straight lines, as though the blade repeatedly replicates its form, creating a pattern that straightens the directions of thoughts. Hearts are another well-known symbol, often used to represent love in superficial ways, such as on clothing, in the media, and in the arts. However, its coronal shape is generated by a similar vibratory signature produced in the area of the human heart chakra when affection is experienced. Additionally, the same creative emanation that collectively produces an apple is intrinsically similar in vibration to what emanates from the region of the heart. These examples are archetypal figures that represent a rather simple symbol, although they are potent on other planes. A symbol is not just a simple representation of an entity, location, or object; it communicates, through symbology, the pathway for that entity to progress or for that object to acquire influence. A symbol embodies a particular frequency, carrying meaning and significance beyond its visual appearance. Despite not displaying a frequency representation to the human eye, it creates the desired frequency. When combined, the frequency of symbols functions similarly to the frequency of elements.

Magical symbols, including pentagrams, voodoo sigils, and grimoire seals, represent holographic objects that can act as conduits to other realms. Activating these symbols can grant access to or influence from spirits during rituals. The activation of them progresses as the magician imbues it with mental magnetism, thereby transforming it into a thought-form and subsequently promoting its magnetic properties. Some priests may trace and link symbols to multidimensional portals indicated by an entity. A sigil, or magic symbol, is a representation of a specific frequency or idea and is often designed with an aesthetic appeal and mathematical outline that represents, in the form of a symbol, the properties needed for that multidimensional channel.

Magical symbols function similarly to national flags, which represent entire nations. Vibrations can be expressed as frequencies, which can be depicted in graphs and translated into numbers. Numbers can be represented as angles and traces, all of which can be contained within a single sigil. The meanings, angles, and words contained in a sketch aim to create an element out of pure vibration. In other words, a symbol can reproduce all the qualities of a crystal, with its mathematical visuals imprinted with those that are frequently observed in crystals. Therefore, a combination of abstract shapes, angles, and traces may be as potent as a combination of material elements if properly activated for their extraphysical purposes. It is worth noting that the symbols must be activated before they can exert any extra-physical function. Groups of spirits can use symbols as conduits for transmitting information to and from the physical world through incarnate individuals. These symbols act as portals through which the spirits can communicate, similar to a walkie-talkie. The communication is neither verbal nor symbolic but telephatic through vibratory fields.

As an example of a talisman, a gris-gris is a small handmade pouch that typically contains a combination of seeds, rocks, hair, or nails, as well as several symbols or objects believed to attract the energies needed for the desired outcome. For instance, a gris-gris created to facilitate pregnancy may contain a small diamond or aquamarine crystal, a kola nut, a handful of lavender, several silver, nickel, or platinum coins, a sphere of clay made from mud collected near a river shore and mixed with menstruation blood, and four shells collected from the river or beach. All the elements included in a talisman like a gris-gris must have a similar vibratory pattern and be created by the frequency of Generation, which is part of Source or God, and is related to the creation of physical and non-physical particles and waves. Despite being sustained by distinct vibrations, both the diamond and the kola nut were generated by similar vibratory states. The diamond, a form of life created by the divine frequency of Generation, is expressed in the mineral kingdom and maintained by the divine frequency of Knowledge. Similarly, the kola nut, generated by the frequency of Generation, is expressed in the plant kingdom and sustained by the frequency of Justice.

When considering God and its creations, it is crucial to remember that God is not a person, a spirit, or an entity, but rather energy, vibration, and frequency beyond the limits of human comprehension. As a compassionate and intelligent force, God brings into existence and maintains the physical universe, resulting in the expression of life through natural forces present in every kingdom, organic and inorganic. Despite being perceived as the sole God, God's

"frequencies" can be grouped into various fields of manifestation, including Evolution, Generation, Justice, Love, Law, Knowledge, and Faith. These fields can be characterized as follows:
- Evolution relates to the transformation of life, maturity, and rebirth.
- Generation relates to the creation of life or the interruption of it.
- Justice relates to the balance of life and accrued merits and karma.
- Love relates to youth, beauty, harmony, and prosperity.
- Law relates to life according to the universal laws, such as cause and effect.
- Knowledge relates to the expansion of life in the direction of enlightenment.
- Faith relates to a belief that extends beyond physical life and space-time.

As these elements carry a frequency mark, they are more likely to attract similar elements that have the same signature. It could be a holographic object, such as a thought, or it could be any other object that can materialize and enter physical reality, as the elements present in the gris-gris act as a condensed magnet for the desired similar counterparts. Along these lines, the elements are materialized versions of symbols, whereas symbols are depictions of elements.

Being carried by the individual, the electromagnetism of these elements will magnetize the individual's aura, eventually altering their vibratory state and bringing them into greater resonance with the wave of Generation. In other words, their capacity to co-create and maintain a fetus is likely to arise since only individuals whose auras resonate at the frequency of Generation are capable of conception, which necessitates hormonal and psychological harmony. Frequently, people seem to wish to have a child; nevertheless, their unconscious mind rejects the possibility since a child would disrupt their lifestyle and compel them to change, which their ego would undoubtedly try to suppress. Hormones are also significant, not just for obvious endocrinologic reasons, but as a physical element with extra-physical equivalents that play a role in permitting similar energy to infiltrate the body.

Like symbols, amulets and talismans must still be activated through ritual. Activation may be accomplished in a variety of ways, most notably by using one's auric field, which serves as fuel that permeates the elements. In shamanic and spiritualist practices, an entity being channeled blesses or magnetizes an amulet, thereby connecting their auric emanations to the object and making it a conduit of their spiritual signature. This can be useful in cases where the individual is developing mediumship and needs to become accustomed to the entity's frequency near their aura. The entity can also impart a certain frequency to the object, making it easier for energy to flow into an altar where it is maintained. The elements in a amulet must consist of similar frequencies for the deity to facilitate the procedure. Spirits can also play a role in the activation of the symbols contained in the talisman, entangling the it to the individual so that the emanations of one favorably affects the other.

A miniature house made of glass, wax, wood, or resin cannot attract a real house to the person who keeps it as an amulet. However, by merging a thought form with the miniature, the object's electromagnetic field can be excited, allowing for photons and etheric magnetism to transfer and thus making it an etheric magnet. In other words, the movement of particles and waves in the miniature will make the miniature a receiver of magnetism. This may enhance its sphere of extra-physical action and increase its potential to attract similar objects, such as a house.

However, it is important to note that the miniature itself does not possess the power to attract a house independently. Rather, it can influence the person to think and act in alignment with the idea of acquiring a house. The thought-form embodied in the miniature can even magnetize the actual house, provided that the individual's energies are aligned with their intention. Overall, the miniature object serves as a bridge between the desire for a house and the frequency of houses.

In conclusion, symbols and sigils play a crucial role in attracting analogous frequencies to objects. They are not just simple representations of entities, locations, or objects but also communicate pathways for extraphysical realities to progress and acquire magnetism, often taking the form of mathematical or archetypal representations.

Oracular Tools

Oracles are divinatory tools that are utilized by both paranormal and ordinary individuals to obtain occult messages and answers from a different plane of existence. These tools serve as a medium for the translation and decoding of hidden information and act as a conduit between intuitive knowledge and conscious awareness. Although there is a plethora of oracles available, it is essential to note that astrology, face reading, and palm reading (not to be confused with palmistry) are not classified as oracles. These methods rely on analytical maps, mathematical or geometric observations, correlations, and analogies to decipher reality. On the other hand, an oracle is defined by the communication that it promotes between auras. In order to gain a better understanding of the mechanics behind the functions of a simple oracle, such as the tarot or other oracular decks of cards, various phenomena studied or theorized by modern physics can be utilized. However, it is important to note that any similarities between oracles and scientific evidence may be purely coincidental.

Firstly, as theorized by Einstein et al. in 1935, quantum entanglement occurs when two sub-atomic particles become linked and cannot be described independently of each other's existence. Scientists have manipulated pairs of molecules in various experiments over the decades, and quantum entanglement has been experimentally demonstrated with photons, neutrinos, electrons, molecules as large as 60 carbon atoms, and even small diamonds (Lee et al., 2011). Furthermore, the Pauli exclusion principle states that no two electrons in an atom can have the same set of four quantum numbers, which includes spin. This means that if one electron has a particular spin, the other electron in the same orbital must have the opposite spin. Experimental evidence has supported this principle, which has been fundamental to the development of modern physics. In fact, in 2018, the European Organization for Nuclear Research (CERN) tested this principle in an experiment and found that electrons behaved exactly as predicted, proving the Pauli exclusion principle to be true (for more information, see an article in the March 2018 issue of the CERN Courier).

In this regard, the interactions between oracle objects and thought-forms can be compared to an entanglement, whereby what is present as crystallized thought is paired with the outcome answers from oracles. As a thought-form is excited by a question, so is a card. However, it is important to emphasize that the aforementioned theories are presented as mere illustrations to help the reader follow a scientific line of thinking.

In the practice of tarot reading, the communication is, in fact, between the querent and the deck's auras. The reader serves as a conduit between the querent and the deck, acting as an interpreter of information as well as providing psychic or intuitive interpretations if needed. In the case of a psychic reader, insights into the consultant's past memories, present thought-forms, and future possibilities, particularly those of a superficial nature, may be gained. At the beginning of a reading, the reader subconsciously connects with the querent, and as the cards are shuffled, displayed, and chosen, the aura of the deck resonates with each object present in the querent's aura. As a mediator between the deck and the querent, the reader unconsciously selects the cards based on the information contained in the querent's aura, therefore choosing the cards that correspond to each thought-form or holographic object. Similarly, before the querent chooses the card, their most prominent thought-form resonates with and thus excites the card that is most similar in frequency to it, leading the reader or the querent to unconsciously choose that very card. This is due to the fact that each card is encoded by archetypes, which are essentially the core preconceived concepts and meanings conveyed by each card.

In the realm of divination, each oracle is associated with and sustained by egregores, which are groups of spirits or emanations from a group of spirits. Therefore, the deck is subject to these energies, and even if the reader has no knowledge of the cards' meanings, the accurate and true answer will still be revealed. The cards' quantum entanglement with the querent enables a small group of cards to precisely expose the information in their auras, despite the reader shuffling and randomly selecting the cards. Each card in a specific deck has a corresponding auric image and energetic signature that are engraved with characteristics by the reader and the universal system governing those cards. This ensures that a particular tarot card always holds the same meaning worldwide, regardless of who reads it. Furthermore, memories, thoughts, or holograms of potential future events correspond to specific cards or groups of cards, and this concept applies to other types of oracles, such as cowrie shells, coins, or runes, besides deck-style oracles. Tarot cards and other oracles typically provide correct answers, with the mostly magnetized "possibility of the future" or "crystallized past" available during the actual reading. Apparent inaccuracies may arise due to changing thoughts, emotions, and actions, as well as an incoherent reader's misreading. However, during the reading, the captured information regards that exact moment and does not count on a possible change of factors thereupon.

Some individuals may believe that observing an oracle reading can influence the results and lead them to choose the cards reflecting their fears or desires rather than accurate responses to their queries. However, observation encompasses what is already crystallized and therefore probable in their own auras. Richard Feynman's double-slit experiment in 1965 demonstrated that the electron's path appeared to change upon "observation," leading to the idea that observation affects reality. Although observation indeed affects reality, the term "observation" is often misunderstood as simply the watching of the experiment, when in fact it refers to the role of sensors located in the gaps rather than a human "observing" the experiment. Therefore, watching an oracle tool will not influence any outcome. Only the unconscious observance of an oracle may impact responses through one's crystallized auric objects.

Palmistry

Palmistry is a method of analyzing the energetic characteristics of the hands, specifically the palms, with the purpose of interpreting events from the past, present, and future. The reader's intuition guides the interpretations, and it is important to consider this aspect when evaluating the accuracy of a reading. Furthermore, it is essential to differentiate palmistry, also known as chiromancy, which relies on intuition, from chirology, which focuses solely on the physical aspects of the hand, such as the length and width of lines, hills, and bumps. While chirology observes the physical characteristics that one's energy imprints on the hands, palmistry directly examines one's etheric energies from the chakras and their canals of traffic, using the palms as a surface where most canals terminate.

According to Motoyama (1981), the energy canals of the body, which extend from the chakras to the organs, tissues, and bodily fluids, are known as energetic meridians or nadis. The energies that travel through these canals have different names in various cultures, such as chi, prana, and vital force, and are absorbed by one's numerous chakras and subchakras. Etheric energy also traffics through these meridians, and this energy is generated by cell movement and respiration and is influenced by one's psyche, or mental and emotional state. However, it is essential to note that prana and etheric energy are not the same force. Prana can be divine, spiritual, or a natural emanation of vitality, whereas etheric energy is the subtle energy generated by organic life. Although these two types of energy are sometimes used interchangeably in esoteric literature, their distinct meanings do not differentiate in terms of the energy canals where both types of energy travel. Moreover, prana invigorates etheric energy, and the two can coexist in the same plane and at the same frequency.

The energies that comprise the etheric body, including

the absorbed prana, are considerably stronger near the major chakras of the body, which act as energy redistributors and converters of energies to and from all meridians and their branches. These canals also derive into main canals alongside the spine and branch into minor canals, to the point where they appear to be smaller than ciliary veins. The canals, or meridians, are similar in presentation to the nervous system, but they do not necessarily share the same pathways, nor are they physical and non-physical counterparts of one another. As stated, a person's vital energies, which have a semi-physical constitution, are generated by various sources, including cosmic waves, ancestral electricity passed from DNA, respiration, diet, and relationships. Depending on the quality of the sources, these energies can have a positive or negative impact on an individual's overall well-being and can be managed through practices such as contact with nature, healthy eating, adequate sleeping hours, laughter, and positive social connections. In like manner, the endocrine system and its seven glands also serve as the primary propellers of these energies in the form hormones, which play a significant role in shaping one's health and feelings, and thus one's semi-physical state.

 An individual's physical, energetic, and spiritual constitution can be evaluated by the condition of their energy vortices and energetic canals, which are visible on various parts of the body, including the palms, soles of the feet, ears, iris, and tongue. The physical state of these regions can offer insights into how and what energies imprint on them, thus telling a story. Although energy readings can be observed on different regions where the canals terminate, such as the iris and the ears, the palms are often the most common and convenient location for assessing energy due to their size. In chirology, which involves the study of the hands' physical aspects, the ends of the meridians can be analyzed similar to other techniques that associate specific body parts and their aspects with health or emotional state, such as acupuncture, reflexology, and Tai Chi Quan. Meanwhile, in palmistry, which uses intuition and sensitivity, readers decode the quality of the energies observed and sensed. It is important to note that only individuals with exceptional sensitivity and psychic abilities can become experts in palmistry. Therefore, palmistry readings require a skilled and intuitive practitioner who can accurately interpret and assess the individual's energetic and spiritual constitution. In contrast, for chirology, theoretical studies and practical experience may suffice to decode the meaning behind certain physical aspects and their corresponding energetic associations.

Overall, palm reading resembles aura reading in all its methods, the only difference being the focus on the emanations of one's field or on the termination spots of all energetic canals. For individuals who are sensitive to energy, the various energies emerging on the palm may reveal a spectrum of information, including colors, fragrances, sensations, and even small pictures of their aura. However, both chirology and palmistry are reflections of the energetic system and can be used to investigate an individual's extra-physical energy system. As a result, they are often used in conjunction with one another to provide a more comprehensive understanding of an individual's energetic makeup.

Supernatural Games

The utilization of amateur methods to communicate with the dead, such as Ouija boards, can yield negative consequences for their participants. The act of invoking a spirit through concentrated intention may create holographic beings instead of attracting genuine entities. The participants' desire for a response can cause their thought-forms to linger around the oracle board or objects used. Although no actual ghost materializes, the electrical discharges from their minds animate these holograms on the etheric planes, which often take the form of a tormented spirit. Although the holographic "spirit" lacks the ability to move objects, such as a pointer or a cup, it may provide feedback as a virtual entity if one of the participants is a medium capable of producing physical effects and externalizing ectoplasm. The thought-form generated during an oracle game may assume a life of its own, albeit without the ability to reason. The psychic, together with other participants connected to the hologram, serves as the driving force behind the virtual ghost's activities. However, without its own energy source, the hologram usually dissipates within a few hours if the players stop maintaining it through energetic transfer. Hence, for such games to work, the presence of a psychic who can produce physical effects is necessary.

It is crucial to take note that these "oracle games" can have spiritual implications for the participants. Some spirits may derive energy from a participant's energy source or develop an obsession while attempting to communicate with the physical world. In many cases where participants employ objects such as pendulums or cups that allegedly move during divination, the apparent response from the oracle can be traced back to the participant's subconscious mind. The autonomic nervous system, which comprises the sympathetic and parasympathetic nervous systems, manages most of the body's functions without conscious will. The sympathetic division is responsible for immediate and conscious reactions, such as the fight-or-flight response, while the parasympathetic division regulates activities that do not require immediate action, such as digestion (For more on the autonomic nervous system, see Jänig, 2013, pp. 179–211). Along these lines, balancing a compass, holding a pendulum, or gliding a cup over a specific answer on the board are immediate responses based on the player's own sensibility. Since the conscious mind cannot easily access extrasensory information, the unconscious mind processes the information cognitively in the form of a question and conveys it through neuromuscular reactions. Although the motor response is subtle, it is sufficient to guide the player's motor coordination to make an object move through apparent answers. Thus, all the gliding cup, balanced compass, and pendulum movements rely on genuine occult information, in which the nervous system transmits subtle information to the object through subtle movement. This movement is not perceived by one's motor senses but reflects the movements that one subconsciously knows.

Overall, in the absence of mediums among the group, participants do not consciously determine the answers but are most likely responding with their own subconscious mind.

BIBLIOGRAPHY

Aftanas, L. I., & Golocheikine, S. A. (2001). Non-linear dynamic complexity of the human EEG during meditation. Neuroscience Letters, 310(1), 57-60.

Anderson, J.E. (2015). The Voodoo Encyclopedia: Magic, Ritual, and Religion. ABC-CLIO.

Avidan, A.Y., & Zee, P.C. (2011). Handbook of Sleep Medicine (2nd ed., Chapter 5). Lippincott Williams & Wilkins.

Barrett, D. (1992). Just how lucid are lucid dreams? Dreaming, 2(4), 221–228. https://doi.org/10.1037/h0094362 [Accessed in December 2018].

Borges, W. (2005). Extraphysical and Projective Teachings Spiritual Guidance by Sanat Khum Maat. Original pt madras.

Brandenburg, A., et al. (2007). Hydromagnetic dynamo theory. Scholarpedia, 2(9), 2309. https://doi.org/10.4249/scholarpedia.2309 [Accessed in January 2018].

Cahn, B. R., & Polich, J. (2006). Meditation states and traits: EEG, ERP, and neuroimaging studies. Psychological Bulletin, 132(2), 180-211.

CERN Courier. (2018). "Exclusion principle. Putting the Pauli exclusion principle on trial." Retrieved from https://cds.cern.ch/record/2315220/files/vol58-issue2-p035-e.pdf. [Accessed in May 2018].

Connors B., Long M., (2004). "Electrical Synapses in the Mammalian Brain," Annual Review of Neuroscience, 27, 393-418.

Cuoco A., (2014). African Narratives of Orishas, Spirits and Other Deities: A Journey Into the Realm of Deities, Spirits, Mysticism, Spiritual Roots and Ancestral Wisdom. Outskirts Press.

Darwin, C. (1872). The Expression of the Emotions in Man and Animals. John Murray, London.

Dirac P., (1947). The Principles of Quantum Mechanics, 2nd edition. Clarendon Press.

Eakin, R. M. (1973). The Third Eye. Berkeley. University of California Press.

Edwards R., et al. (1990). "Measurements of human bioluminescence," J Biolumin Chemilumin, 5(4), 205-7.

Einstein, A. (1905). Does the inertia of a body depend upon its energy-content?. Annalen der Physik, 18(13), 639-641.

Einstein, A. (1925). Unified Field Theory of Gravitation and Electricity. Session Report of the Prussian Academy of Sciences, 414-419, July 25th.

Einstein A., Podolsky B., and Rosen B. (1935). "Can Quantum-Mechanical Description of Physical Reality Be Considered Complete?," Phys. Rev. 47, 777. Institute for Advanced Study, Princeton, New Jersey.

Einstein A., Infeld L., (1938). The Evolution of Physics: The Growth of Ideas from Early Concepts to Relativity and Quanta. Cambridge University Press.

Feynman, R. P., Leighton, R. B., & Sands, M. (1965). The Feynman Lectures on Physics, Vol. 3. Addison-Wesley.

Gagnon, P. (2011). Pauline Gagnon's Blog: Did we build the LHC just to find the Higgs? (No. BUL-NA-2011-265). CERN Publications. IR-ECO-CO. Retrieved from https://cds.cern.ch/record/1387916?ln=en [Accessed in March and April 2018].

Griffiths, David J. (2005). Introduction to Quantum Mechanics, 2nd edition. Pearson Prentice Hall.

Groner, J.P. (2017). The Human Body as an Electric Generator. IEEE Pulse, 8(2), 10-12.

Hodson, G. (1952). The Kingdom of the Gods. Theosophical Publishing House.

Jänig, W. (2013). "The Autonomic Nervous System." In C. Galizia & P.M. Lledo (Eds.), Neurosciences - From Molecule to Behavior: a university textbook. Springer Spektrum.

Jinarajadasa, C. (1921). First Principles of Theosophy. Theosophical Publishing House.

Koenig, L. B., et al. (2005). Genetic and environmental influences on religiousness: Findings for retrospective and current religiousness ratings. Journal of Personality, 73(2), 471-488.

Korotkov K., et al. (2004). (Ed.). Measuring Energy Fields: Current Research. Backbone Publishing Co. Fair Lawn, USA, 2004. pp.157-170.

Lang, S., & Baconnier, S. (2002). Electric field interactions in mitosis: A review of the evidence. Journal of Cell Biology, 158(4), 389-396.

Lee, K. C., et al. (2011). "Entangling macroscopic diamonds at room temperature." Science, 334(6060), 1253–1256.

Lemaître, G. (1927). Un Univers homogène de masse constante et de rayon croissant rendant compte de la vitesse radiale des nébuleuses extra-galactiques. Annales de la Société Scientifique de Bruxelles, A47, 49-59.

Masaki Kobayashi, Daisuke Kikuchi, and Hitoshi Okamura (2009). "Imaging of Ultraweak Spontaneous Photon Emission from Human Body Displaying Diurnal Rhythm," PLoS One, 4(7), e6256.

Moorthy, C. S. (2011). Gleanings from Rig Veda: When Science Was Religion. Authorhouse.

Motoyama, H. (1981). Theory of the Chakras. The Theosophical Publishing House in Wheaton, Illinois.

NHS, (n.d.) https://www.nhs.uk/conditions/sleep-paralysis/#:~:text=Causes %20of%20sleep%20paralysis&text=insomnia,%2Dtraumatic %20stress%20disorder%20(PTSD) [Accessed in December 2018].

Pereda, A.E. (2014). Electrical synapses and their functional interactions with chemical synapses. Nature Reviews Neuroscience, 15, 250-263. https://doi.org/10.1038/nrn3708

Philosophical Research Society. (1958). Philosophical Research Journal. Volumes 18-19.

Piaget, J. (1967). The Child's Conception of Physical Causality. (Original work published in French in 1926). New York: Harcourt Brace Jovanovich.

Reiter, R. J., et al. (2016). Melatonin as an antioxidant: Under promises but over delivers. Journal of Pineal Research, 61(3), 253-278.
Reiter, R. (2019). Melatonin and the Optics of the Human Body. Medical Research Archives, 7(11), 1-8.

Richter-Levin, G., & Akirav, I. (2000). Amygdala-hippocampus dynamic interaction in relation to memory. Molecular Neurobiology, 22(1-3), 11-20.

Rolls, E.T., & Grabenhorst, F. (2008). The Orbitofrontal Cortex and Beyond: From Affect to Decision-Making. Progress in Neurobiology, 86(3), 216-244.

Rosen, C., et al. (1992). Piezoelectricity. Key Papers in Physics. American Institute of Physics.

Rubik, B., et al. (2015). Biofield Science and Healing: History, Terminology, and Concepts. Global Advances in Health and Medicine, 4(Suppl), 8–14.

Russell, C. T. (1972). The configuration of the magnetosphere. In E. R. Dyer (Ed.), Critical problems of magnetospheric physics: Proceedings of the Symposium held 11-13 May, 1972 in Madrid, Spain (pp. 1-13). National Academy of Sciences.

Seligman, M. E. P., & Yellen, A. (1987). What is a Dream? Behaviour Research and Therapy, 25(1), 1-24.

Sharpless, B.A. (2016). A clinician's guide to recurrent isolated sleep paralysis. Neuropsychiatric Disease and Treatment, 12, 1761-1767.

Sternberg, R. J. (1999). Cognitive psychology (2nd ed.). Wadsworth Publishing.

The Oxford Dictionary of English (2nd ed.). (2003). Oxford University Press.

Von Bartheld, C. S., Bahney, J., & Herculano-Houzel, S. (2016). The search for true numbers of neurons and glial cells in the human brain: A review of 150 years of cell counting. The Journal of Comparative Neurology, 524(18), 3865-3895.

Zhao, Y., & Zhan, Q. (2012). Electric fields generated by synchronized oscillations of microtubules, centrosomes and chromosomes regulate the dynamics of mitosis and meiosis. Theoretical Biology and Medical Modelling, 9(1), 26. https://doi.org/10.1186/1742-4682-9-26

RECOMMENDED READINGS

The Spirits Book by Allan Kardec (1857)

Thought-Forms by Annie Besant and C.W. Leadbeater (1905)

Missionaries of the Light by Francisco Cândido Xavier (1945)

Autobiography of a Yogi by Paramahansa Yogananda (1946)

The Kingdom of the Gods by Geoffrey Hodson (1953)

Spiritual Travel: The Projection of Consciousness by Wagner Borges (2017)

Other Books by the Author

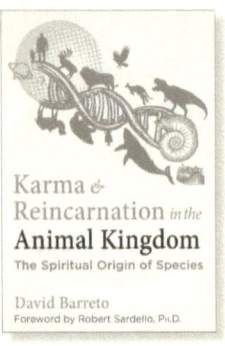

Karma & Reincarnation in the Animal Kingdom
The Spiritual Origin of Species

Explores the spiritual nature of animals, their reincarnation and karma processes, psychic experiences within species, and the spiritual consequences of meat consumption.

The Book of Umbanda
The Grand Compendium

A comprehensive reference for scholars and devotees alike on the largest folk religion in the Americas, exploring its history, rituals, and the diverse array of worker spirits.

One Hundred Messages From Above
For a Rapid Spiritual Ascension

A collection of 100 messages to accelerate one's path to spiritual ascension. These kind - but firm - advice covers both challenging and taboo themes: from mental illnesses, suicide, abortion, and addiction to resilience, past lives, ageing, and fame.

www.ingramcontent.com/pod-product-compliance
Lightning Source LLC
Chambersburg PA
CBHW031121080526
44587CB00011B/1061